BIOENERGY
HEALING

BIOENERGY HEALING

An alternative approach
to cancer and other diseases

My journey as a healer using the
POWER *of the*
UNIVERSAL ENERGY FIELD

Learn the Foundation of Bioenergetic Healing
and Experience Vibrant Health Using Simple
Natural Techniques

ANUSHAVAN MANUKYAN

Copyright © 2015 by Anushavan Manukyan.

Library of Congress Control Number:		2013908013
ISBN:	Hardcover	978-1-4836-3572-9
	Softcover	978-1-4836-3571-2
	eBook	978-1-4836-3573-6

All rights reserved. No part of this book may be reproduced or transmitted in any form or by any means, electronic or mechanical, including photocopying, recording, or by any information storage and retrieval system, without permission in writing from the copyright owner.

Print information available on the last page.

Please note: The information provided in this book a compilation of events and personal experiences that have NOT been scientifically verified or corroborated by medical experts. It is anecdotal information and should be treated as such. For serious medical concerns please consult a qualified health care practitioner.

Rev. date: 11/15/2018

To order additional copies of this book, contact:
Xlibris
1-888-795-4274
www.Xlibris.com
Orders@Xlibris.com
546320

Cancer?
Is it Even Easier to Cure than Muscle Pain?

We all have the capacity to heal using the power of the bioenergy field

In 1972, Anushavan Manukyan graduated as a quantum physicist from Armenian University in Yerevan, Armenia. In 1982, he began his practice as a bioenergy healer in Armenia, where both adults and children from distant places sought his healing abilities for many serious conditions.

Manukyan collaborated with medical doctors to confirm diagnoses and also performed many miraculous healings, where other methods had failed—using only the power of bioenergy. However, until his wife received a diagnosis of breast cancer, he had avoided using these techniques on cancer patients, as there was a belief that the healer could be affected . by the cancer. Compelled to save his wife, Manukyan broke through the perceived barrier and discovered that he could safely use bioenergy therapy to eliminate many disease states , including cancer.

Manukyan immigrated to the US, and in 1988 became the first licensed "bioenergy healer," with offices in Hollywood and Beverly Hills.

This universal energy field is available to all of us. The author's dream is for all people to know that they have healing abilities that go beyond what medical institutions can offer.

His book encourages the reader to find their own healing path and master their innate self-healing abilities, as he did. We can all learn basic foundational approaches to a healthy lifestyle and how to use the power of the bioenergy field to support our mind, body, soul and our journey through life.

Anushavan Manukyan practices and teaches bioenergy healing in Los Angeles. He can be contacted at info@HealingWithBioenergy.com

Contents

Preface 13

Author's Note 15

Acknowledgments 17

Introduction 19

Chapter 1: So What Is Bioenergy? Is That the Same as an Aura? 21
    The Kirlian Effect 22
    They Are Even Talking about It on the Popular American Television Program, Good Morning America! 24
    Tumors—Many Are Simply an Accumulation of Negative Energy 27
    Soviet Scientific Research: Meditative Access to and Manipulation of the Biofield 28
    Seeing the Evidence of Bioenergy on Radiological Scans 29

Chapter 2: The Journey of How I Developed My Ability to Heal 31
    My First Exposure to Hypnosis 32
    My Interest in Hypnosis Is Rekindled 33
    My Introduction to Yoga 35
    Learning to Heal 36
    Meeting a Mentor in the Study of Bioenergy Healing 38

Chapter 3: My Wife Elada's Breast Cancer 41
    It's OK to Heal Everything—Except Cancer 41
    Coming Face-to-Face with Cancer 41
    Protecting Myself 43
    Healing Cases 46
    After Years of Chronic Kidney Pain, Albert Now "Treats Himself" Using Bioenergy! 47
    A Three-Year-Old "Destined" for Institutionalization 48

Anna Has Unexplained Neurological Disorders *49*
A Breast Tumor That Disappeared in 10 Minutes! *52*
Hypnosis and Bioenergy Therapy—A Powerful Combination *54*
A Four-Year-Old Boy with a Malignant Neck Tumor *56*
Seda Recovered from Terminal Lymphatic Cancer—
Her Doctor Drops His Glasses! *58*
We Learned of Tomas's Discharge from the Army with an
Adrenal Tumor and Anush's Miraculous Healing! *62*
The Doctors Said, "We've Never Seen Lupus-Related Osteoporosis
Reverse Itself!" *64*

Chapter 4: Bioenergy Begins to Capture the Attention of the
Soviet Government and Scientists 67
"Visitors" from Moscow *67*
Validation by the Scientific Community in Yerevan, Armenia *69*
Requests from the Medical Community to Diagnose
Hospital Patients *70*
Collaborating with and Inspiring Professor Amatuny to
Add Bioenergy to His Medicine Bag *73*
Armenian Government Officials Seek Healing *75*

Chapter 5: Our Plans to Move to America: My Healing Art
Paves the Way 79
An Appreciative Father Brings Us Closer to America! *79*
Healings on Arbat Street *81*

Chapter 6: Establishing My Healing Practice in the United States 85
Absorbing the Benefits of Someone Else's Bioenergy Treatment! *86*
Liver Disorders, Including Cancer, Are Very Responsive to
Bioenergy Therapy! *87*
There Are No Guarantees in Life—But Bioenergy Gives You Very
Favorable Odds! *88*
Bioenergy Bolsters the Immune System—
Elimination of Breast Cancer *90*
Prostate Cancer Patient Enlists My Help for a Remote
"Emergency" Healing *91*
Even Celebrities Were Hearing of My Work! *93*
At 19, Jennifer in "an irreversible coma" *96*
We Are More than Just This Physical Body *99*
Three Cases—One Family *100*

Chapter 7: Cancer—Does the Medical Establishment
Need a New Approach? 113
   Strides in Cancer Treatment 113
   Another Perspective to Consider 114
   Psychological Impacts of a Cancer Diagnosis 115
   Current Approaches to Cancer Treatment 115
   Why Surgical Intervention Is Frequently Counterproductive 117
   Treatment Using Bioenergy. 118
   A New Approach to Healing. 120
   The Four Components of the Human Body 121
      1. *The Physical Body* 121
      2. *The Ethereal Body* 121
      3. *The Astral Body*. 121
      4. *The Mental Body*. 121
   Simple Reminders about Your Healing Capacity 122

Epilogue. 125

Appendix A: Creating a Healthy State Using the
Power of the Bioenergy Field 131

Exercises. 133
   Overview 133
   Concentration and Willpower Training 134
   More Willpower Training. 134
      Exercise: Mirror Training. 135

Perceiving Your Biofield 137
      The following exercises will help you to learn to perceive and
         ultimately manipulate your biofield.
      Preparation—Pharaoh Position 137
      Exercise: Sensing the Biofield with Your Palms and Fingertips. . . 137
   Strengthening the Sensitivity of Hands. 138
      Exercise: Sensing Objects in the Biofield (using one hand). 138
      Exercise: Sensing Objects in the Biofield (using both hands) 139
      Exercise: Creating a Line of Energy. 139
   Programming the Biofield 140
      Exercise: Creating Cosmic Energy (Bioenergy) Balls 141
      Exercise: Moving the Energy Ball 141

- Exercise: Working with Multiple Energy Balls 142
  - Part 1 142
  - Part 2 142
  - Part 3 142
  - Part 4 143
- Controlling and Programming Your Biofield 143
  - Exercise: Flower and Seed Growth Visualization 143
    - Part 1 144
    - Part 2 144
    - Part 3 144
- Working with Energetic Centers or Chakras. 145
- Sensing Energy Flow and the Location of the Chakras 147
  - Sensing the Energy Flow. 147
  - Sensing the Chakras. 147
    - Part 1 147
    - Part 2 148

Appendix B: Getting Well . . . Keeping Well 149
- Enhanced Breathing. 150
- Breathing Experiment and Introduction to "Full Breathing". 150
- Water—The Elixir of Life. 151
- Water's Molecular Structure Responds to Our Energy 151
- Water Consumption Requirements 155
- Finding a Clean Water Source 156
- Oxygen-Supplemented Water—Added Benefits! 156
- Again a Little about Water. Did You Know . . . ? 157
- My Old-World Diet. 158
- My Breakfast. 158
- Lunch Tips 159
- Dinner and Yogurt. 159
- Maintaining Prana or Life Energy, Also Referred to as Gold Energy, Cosmic Energy . . . Bioenergy 160
- Physical Health and Yoga 161
  - The Beginning of Every Day: A Routine to Follow 163
    - Upon Waking 163
    - Upon Rising 163
  - Opening and Loosening Your Joints. 166
    - Ankle Rotations 166

    Knee Rotations................................... 166
    Forward Leg Swings.............................. 166
    Side Leg Lifts................................... 167
    Hand Stretch.................................... 167
    Wrist Rotations................................. 167
    Wrist Bend (Up and Down)........................ 167
    Wrist Bend (Left and Right)..................... 167
    Forearm Rotation (Forward and Backward)......... 167
    Arm Rotation (Forward and Backward)............. 168
    Neck Stretch (Forward—Backward)................. 168
    Neck Stretch (Left—Right)....................... 168
    Neck Rotation (Left—Right)...................... 168
    Forward Neck Roll............................... 168
Pranayama—Yoga Breathing Techniques................. *169*
Prepranayama: Learning How to Fully Breathe!........ *170*
    Abdominal Breathing............................. *170*
    Midlung Breathing.............................. *170*
    Chest Breathing................................. *171*
    Full Breathing.................................. *171*
Pranayama Exercises................................. *173*
    Opening lung cells............................ 173
    Cleansing breathing........................... 174
    Broadening of lungs........................... 174
    Squeezing of the lungs........................ 175
    Nerve calming exercise........................ 175
    Isolated abdominal breathing.................. 176
    Breathing in the "lion position".............. 176
    "Moon and sun" breathing...................... 177
Asanas.. *178*
    Shavasana-Asana or Corpse Pose................ 179
    Morning Asanas................................ 181
        A. *Utthan Padasana—The Leg Lifting Pose.*....... *182*
        B. *Pashimotanasana—The Noble or Powerful Pose*.. *182*
        C. *Bhujangasana—The Cobra Pose.*................ *183*
        D. *Salabhasana—The Locust Pose.*................ *183*
        E. *Sarvangasana—The Shoulder Stand Pose*........ *183*
        F. *Matsyasana—The Fish Pose.*................... *184*
        G. *Trikonasana—The Triangle Pose*............... *184*

  H. *Dhanurasana—The Bow Pose* *184*
  I. *Halasana—The Plow Pose*. *185*
 Evening Asa. 185
 Spiritual Cleansing of the Chakras 188

Chapter8: More of the Science That Supports the
Concept of Bioenergy. 191
 The Evolution of the Fundamentals of Physics Provides
  Evidence of Bioenergy 191
 Aura and Energy Field Hypotheses. 193

Index 195

Preface

A diagnosis or even a suspicion of cancer is one of the most dreaded news an individual can imagine. Widely publicized statistics echo that facing cancer is virtually inevitable in our lifetime or in the lives of those most dear to us. In writing this book, I'd like to forever change or at least present a compelling case for you to reconsider several widely held ideas about the nature of cancer and the real mechanisms that will eradicate it—approaches that do not include chemotherapy, radiation therapy, surgery, bone marrow transplants, or even the often more palatable nontoxic treatments, such as enzymes, vaccines, homeopathy, hyperthermia, nutritional supplements, and the list goes on! So what else is there, you may ask?

I would have had the same question some years ago. A physicist by training, well grounded in science and empirical techniques, the idea of trying to utilize the body's "energy fields" as a remedy for disease—these unseen, largely unstudied, and not well-understood phenomenon—would have been an embarrassment to my reputation in the scientific community, to say the least. I would have never imagined the course of events that has led me to where I stand today—to tell you that there are ways of healing that have always existed—and are the root of how we are able to heal from cuts, burns, surgery, and other damages to the integrity of our bodies. These healing abilities that we all have are underutilized but, when harnessed, can bring about rapid resolution of even cancer in a matter of days and sometimes only minutes or hours.

This book describes to you my journey to mastering the healing skills that ultimately led to my ability to diagnose and eliminate a wide range of diseases that were presented to me, including various neurological, muscular, digestive, and endocrine disorders, and culminating with many forms of cancer.

Along with my narration of my many remarkable opportunities to help individuals on their healing path, wherever possible we have included testimonials.

It is my hope that individuals with cancer, other imbalances, and all of us who take good health as a gift to be protected will adopt the techniques that I intend to share, thus bringing about a new paradigm for the treatment of disease and the ongoing maintenance of a healthy balance in our bodies.

Author's Note

All efforts have been made to accurately convey the details of an individual's medical condition and results of treatment. In some instances, we were unable to contact the parties involved to get their review of their respective stories. Thus, we welcome feedback, corrections or additions to this book at any time.

Please note that in some cases, individuals' names have been substituted with fictitious names shown in italics.

Acknowledgments

I would like to extend my thanks to te following individuals for their generous contributions of time and expertise to make my dream of bringing this book to you a reality: To Kina Merdinian, who reorganized, rewrote, and refined the manuscript as well as researched additional material to take this final product to a new level . Her friendship and support in this endeavor were a blessing and a gift to me . Also to my friend "Masha" from Russia, who is now a Los Angeles teacher, who was kind enough to provide a preliminary translation of my earliest draft from Russian to English; and to journalist huck Schmelter, who provided editing support to that first manuscript . And most importantly to my dear wife Elada, and my daughters, Sona and Arevik, who were by my side after the automobile accident and helped me remember and restore my life's memories and relearn many basic life skills . Because of their love, devotion, and faith in my personal healing capacity, I am now alive and awakened   able to deliver this enlightening book to you and your fam

Introduction

Unfortunately, we have come to know that cancer's devastating effects strike all families to one degree or another—mine was no exception. In 1973, my dear wife, Elada, was diagnosed with cancer. A large tumor was found in her left breast. Doctors determined that it was malignant and recommended surgery. After a great deal of deliberation, we decided to delay taking any steps. Instead, she continued to live with the tumor. In 1980, I had begun to study and further develop my abilities to promote healing in other individuals using *bioenergy*—the human energy field. Eight years after her diagnosis, I had mastered this technique to the degree that I could eliminate a myriad of disease states. After many years of heeding the warnings to avoid attempting to treat patients with cancer, I put the cautions aside and removed the tumor completely from Elada's body.

Over 40 years have passed since then, and she is cancer-free. I will share the details of her recovery and of many of my other patients. As a side note regarding their remissions, often the medical establishment will reverse its position about a cancer diagnosis when the patient is "miraculously cured" by a "healer" or by some "unscientific" modality. Suddenly, tests, images, and diagnostic reports are considered to have been in error. Wouldn't it be nice to have the bills associated with these "now-erroneous" tests and treatments reimbursed as a result! These patients arrived at my door often carrying their medical files of distressing news. Although I was not in a position to validate their condition, it is statistically unlikely that all of them were simply misdiagnosed.

I will describe my journey as a student of several mind-body modalities that ultimately led me to develop these powerful healing abilities, which you can learn as well! Elada too became a student of bioenergy techniques and, out of necessity, miraculously came to my rescue after an automobile accident that left me in a coma. I will return to both of our healing stories later in this book.

Another goal I have in writing this book is to provide insight into the self-healing capacity of our bodies. Our bodies, with all their sophistication, are highly efficient and magnificently well-designed "machines." Our complex structure is capable of performing strenuous physical work, experiencing powerful emotions, and demonstrating deep intellectual reasoning. All of this diverse capability is "computer" driven. When our bodies experience a "malfunction," they automatically send an alert to "the central computer," the greatest minicomputer on earth—the brain—which takes action. Most astonishing of all, this incredible machine—our body—can detect a problem and then actually repair itself. It doesn't necessarily need the help of the "computer technician" or physician as most of us have been taught to believe.

It would take an extraordinary amount of insight to fully comprehend how our bodies are able to perform all of these miracles. What we do know is that the piece of that puzzle that is probably the most critical yet little understood is that bioenergy is a key component in that repair process. Unfortunately for us, our wonderful machines don't come with a handbook or an instruction manual presented at birth. Nor are our self-directed healing abilities widely accepted as real or as a technique to be passed on through parenting or any formal educational process. My hope is that this book will serve as your guide to understanding the bioenergy that sustains you and the self-healing capacity that you have as your birthright.

Thanks to Anush and bioenergy healing, the dangerous mass obstructing my throat and the associated pain and debilitation that my doctors were unable to treat, completely disappeared. I regained my health without drugs or invasive treatments.

Armine Zargarian, Los Angeles, California

CHAPTER 1

So What Is Bioenergy? Is That the Same as an Aura?

There are many forms of energy known to science. Some are clearly understood and even verifiable through reproducible experimentation. Others exist; however, our knowledge is insufficient to explain the entire process of how that form of energy is produced and how it affects the environment. Just as we can feel, observe, and measure the influence of gravity on other objects, explaining the complex principles involved is an entirely different matter. The scientific explanation of bioenergy is similarly difficult to convey. What is agreed upon is that the primary difference between various forms of energy is the length of the energy wave and the frequency and the fluctuations of those waves. You may already know that in many cases one form of energy can be transformed into another. For example, electrical energy can produce the heat in your house, or if you were awake in your physics class, you may remember that potential energy can be transformed into kinetic energy. Consider that the entire universe could be described as a demonstration or display of many different forms of energy.

Bioenergy is present in all living organisms and all living things, including plants, all of which are supported and are surrounded by a bioenergy field—the biofield. You may have also heard the synonymous term *aura*.

We are still far from understanding and explaining the aura phenomenon. The term *aura* or *biofield* has been described as "an invisible light and radiating emanation, which, like a cloud, surrounds living forms and becomes visible to people with open psychic and spiritual vision."

Stories of "auras" or "shells" fill the literature. Drawings depicting these biofields surrounding living creatures have been discovered by unknown artists of ancient civilizations. The famous American psychologist A. Garrett, in a book entitled *Understanding,* writes, "I always saw all the plants, animals, and people surrounded by a mystical shell. In people, it changes in color and stability depending on the mood."

The Kirlian Effect

Although few individuals can see or sense the existence of biofield or aura without training to develop this ability, it is possible to get a photograph of this biofield with the help of the Kirlian effect, or Kirlian photography, named after Semyon Kirlian. In 1939, in the city of Krasnodar, a part of the former Soviet Union, Kirlian produced a photograph of biofield. He accomplished this accidentally. Kirlian, who, as a hobby, tinkered around with various inventions, was intrigued by his observation of the sparks between a patient's skin and a physiotherapy (electronic muscle stimulator) machine's electrodes, when the device was being used therapeutically. (The physiotherapy machine is intended to generate electrical frequencies to control pain and promote healing in a physical therapy setting.) Kirlian observed that the sparks seemed to "dance and jump." Curious, he wondered what would happen if this action was photographed. While holding the electrode connected to a high-frequency physiotherapy machine with one hand and holding an unexposed film and the second electrode simultaneously with the other, he created a connection with his body. Then he decided he would power on the device to see "what would happen!" You knew what was coming and already ran out of the room. Pow! He burned himself! However, the amazing part was that on the exposed film, he discovered an image of his hand surrounded by an unknown form of a visual emanation. It represented what we know today to be a "human aura."

What is now known as Kirlian photography does not actually involve a photographic lens or camera. The apparatus to produce a Kirlian photograph consists of a high-voltage electrical source that is attached to a metal plate. A glass plate sits on top of the metal plate, and a piece of photographic paper is placed on top of the glass plate. The object being photographed, such as a hand or foot, is placed directly on the photographic paper. A Kirlian photograph emerges, consisting of jagged, colored lines that outline the shape of the photographed object. The resulting image is said to represent an aura or an outline of the body's life force.

It was not until the early 1960s that this accident piqued more interest among the scientific community, including the Soviet government. They quickly provided Kirlian with a new apartment, a pension, and a laboratory in which to further his investigations and testing. From that point forward, considerable research and testing began in many scientific research institutions and laboratories across the former Soviet Union. This research sought to answer questions, such as "is there an astral or energy body, as Kirlian's photo seems to imply, and if so, is this astral body repeating or mirroring our physical body?"

At the Russian University in Alma-Ata, a group of biologists, biochemists, and biophysicists restaged and reproduced the Kirlian experiment, so that the Kirlian effect might be viewed through a high-powered electron microscope. With this increased visibility, they were able to actually watch the "live double" and its live organism in action. These experiments were conducted on living plants, animals, and people.

In 1968, Russian doctors V. Inushin, V. Gristchenco, N. Vorobiov, N. Shuysky, N. Feodorva, and F. Gibadullin published their discoveries related to this subject. They announced that all living things—plants, animals, and people—have both physical bodies, consisting of atoms and molecules, and energetic bodies, which resemble and are closely interconnected with the physical bodies. They called it the *plasma biological body*. With the help of Kirlian films produced by Russian scientists, a very compelling additional detail was discovered. After taking a Kirlian photograph of a plant, the scientists cut off one of the leaves and then photographed it again. The photograph revealed an unexpected result. When cutting off a part of the leaf (its physical body), there was no effect on the plasma biological body, i.e., the leaf's outline or "double" remained whole and unchanged.

This was known as the phantom leaf experiment. The same observation was noted in animals and people. Some speculate that this may explain why, following an amputation, some individuals continue to experience sensations, as if the amputated limb were still present.

I witnessed a fascinating version of this phantom limb phenomenon. While tending to some business in a local hospital, a man walked past me, then paused to bend over and scratch his left leg, just below his knee. Initially, I didn't think much of this insignificant event. Coincidentally, ten minutes later, I saw him again—seated in the snack bar enjoying a sandwich. This time his pants were hiked up a bit as he sat in the chair. My gaze was immediately drawn back to that "itchy left leg." To my surprise and wonder, he had no lower leg. Instead, there was a wooden prosthesis there!

They Are Even Talking about It on the Popular American Television Program, Good Morning America!

At the time I began writing this book, there had been an interesting report by *Good Morning America*'s science editor Michael Guillen that scientists are taking a closer look at this concept of auras, which had been dismissed in some Western medical circles. Perhaps the most significant indicator of that recent attention was that the National Institutes of Health in Bethesda, Maryland, decided to start funding research on human *biofields*, our technical term for auras.

Meanwhile, other impressive scientific mavericks were exploring a controversial new science, "bioelectromagnetic." One scientist leading the charge is Berkeley-trained biophysicist Beverly Rubik. She believes that the human body exudes energy, just like glowing candles. *"If you think really what life is: When you're dead, what happens to your energy? Absolutely everything,"* Rubik said. *"You go from being a warm radiant moving creature to a cold dead stiff creature. The big difference is the energy."*

More from the Good Morning America Report (paraphrased)

The human body emits a spectrum of energy like the sun, that can even "broadcast" your state of health!!

Harvard-trained psychologist Gary Schwartz runs the Human Energy Systems Lab at the University of Arizona and believes that the existence of a human aura is like the sun, in that it emits a whole spectrum of energy, both visible light, infrared, ultraviolet, x-rays, gamma rays. He believes the human body is also emitting this whole range of signals, Schwartz demonstrated this theory with Michael Guillen as the subject, sitting close to an antenna that detects low-frequency radiation. By moving his hand close to the antenna, the movement created a signal change on the monitor. An additional high frequency detection antenna responded with a sound as he got closer to it, as it was effectively picking up the "signals his body was broadcasting."

"

The 100 trillion cells inside the body are actually doing the broadcasting, of an aura of electromagnetic radiation, similar to how a television tower functions.

Many mainstream scientists believe this transmission of frequencies is incoherent and meaningless. But other more forward thinking scientists remind us that medical tests such as EEGs and EKGs, measure the electromagnetic signals from the brain and heart, thus it is entirely possible that the rest of the body could also be sending signals about our wellbeing.

Beverly Rubik, who heads the Institute for Frontier Research in Oakland, California, decided to test his hypothesis and Guillen volunteered as a subject. Digital Kirlian photography, widely used in Russia and Eastern Europe, was used to take photos of Guillen's fingertips. The fingertips were chosen because according to Chinese medicine, the fingers contain many acupuncture points, which run on energy channels called meridians to all organs and systems. From the auras of his fingertips, Frontier Research's computer was able to calculate Guillen's complete body aura. Next, it analyzed his aura for anything out of balance or indications of a possible illness. A jagged red circle on the report was an indicator of his state of wellbeing. Rubik's analysis indicated that he had very good energy regulation, however ideal would have been a perfectly uniform circle.

Significant peaks in the jagged red circle were a sign of an excess or inflamed condition. Significant dips implied deficiencies.

The results showed a little bit of "excess" in Guillen's colon region, as well as a couple of organ systems that appeared depleted, including a dip in the endocrine system pointing to a possible deficiency. Guillen corroborated that some of the findings fit with his medical records of an underactive thyroid which paralleled the endocrine system finding of Rubik's detection system.

The National Institutes of Health, also taking the position to research auras, was a sign that progress is being made. We can move beyond today's physics and understand that there could be measurable energies elsewhere in the universe including being broadcast by our cells and about ourselves!

As even the science editor of the *Good Morning America* television show has presented, the movement and, conversely, the stagnation of the bioenergy within our bodies or the body of another human being can be detected. To elaborate on this concept further, these specific locations where energy channels are blocked (also known as accumulations of negative energy), can be sensed using a number of techniques. If not remedied, these energetic blockages can eventually progress, leading to the development of actual physical damage to the surrounding tissue.

Ancient Historic Evidence of Bioenergy

Understanding that if blockages of bioenergy produce disease, then consequently promoting the free flow of bioenergy can restore health. This type of treatment using bioenergy is nothing new and has been practiced since the dawn of time with countless documented examples of its use. I believe the saints, prophets, lamas, dervishes, Sufis, various high priests, and many healers mastered the technique. It was certainly demonstrated by Jesus Christ and Buddha.

Much later, in the nineteenth century, Franz Anton Mesmer, a German doctor, who was a known pioneer of hypnosis and psychoanalysis in medicine, created his method of treatment by influencing the biofield. You may have remembered hearing the term *animal magnetism*. Some of his beliefs, which reference a "fluid," are another way of describing bioenergy and its movement through the biofield:

1. *There is a subtle physical fluid that fills the universe and serves as a medium between the heavenly bodies and Earth, between man and Earth, and between man and man.*
2. *Illnesses arise from the unequal distribution of this fluid in the human body, and recovery occurs when equilibrium is restored.*
3. *With certain techniques (such as the manipulation of magnets around various parts of the body), the fluid can be channeled to different parts of the body and to other persons.*

We have all instinctively used the biofield in an attempt to heal. Think about a time where you experienced pain. We automatically place our hands on or rub an area that is producing discomfort, often until it subsides. A mother, in an attempt to soothe her child, places her hand gently on her child's forehead, and he calms down and goes to sleep.

The biofield can be seen and can also be sensed by the hands. With practice, many individuals can become very sensitive and attuned to even subtle changes in the density and quality of the biofield that can translate to manifestations of an individual's health. One can equate this to the high level of sensitivity that the blind can develop in their fingertips that allows them to read Braille—something that is virtually impossible for the average person. The development of this sensitivity to the ebbs and flows of bioenergy is the basis for how its manipulation can ultimately promote healing.

Later, in appendix A, "Creating a Healthy State Using the Power of the Bioenergy Field," I provide exercises that will introduce you to the basic skills of being able to perceive your biofield and that of another individual.

Tumors—Many Are Simply an Accumulation of Negative Energy

I have observed that the vast majority of tumors will actually yield as a result of the first healing session. How is this possible? My belief is that this is an indication that the tumor itself is somehow more of a potentially transient negative energetic formation than a permanent physical condition. I have come to this conclusion because if it were a purely physical, truly solid formation, it would not typically change its size and density at such a rapid pace. An accumulation of energy, however, could dissipate in a few minutes and, in some instances, even fully disappear. Furthermore, the fact that these negative energy formations are so immediately responsive to bioenergy treatments, makes it also possible for them to reaccumulate at a similarly rapid pace.

Once you practice some of the bioenergy tools and techniques outlined in this book, you may be able to sense these negative energies and formations and clear them. You can practice this yourself with dissolving a hardened mass, such as a tumor. Simply place two or three fingers over the surface and, in a meditative state, imagine it is a piece of ice and you are melting it with the energy of your fingertips. Soon it will gradually soften and shrink—sometimes immediately, sometimes within hours or days.

I have observed that some categories of tumors are completely different—some disappear once and for all and others are more resilient, displaying no evidence of a change during an initial bioenergy session. Yet with subsequent treatments, they shrink or disappear quickly. One patient, whose visible breast tumor had vanished after an initial treatment, returned to me the following day as it had rematerialized. Some might be eliminated again, and then may resurface the next day. This could happen multiple times, until eventually, after repeated bioenergetic treatments, they would disappear permanently. This adds further evidence to the notion that tumors are formed through an accumulation of negative energy. Left unchecked, this stagnated energy can lead to more physical tissue changes, which are more resilient and require repeated intervention to resolve.

Soviet Scientific Research: Meditative Access to and Manipulation of the Biofield

Scientists in the former Soviet Union began researching the existence and behavior of the biofield shortly after WWII. Many notable physicists and mathematicians were active in this area of study. In one compelling experiment, it was demonstrated that a healer or psychic, who could harness the biofield, was able to surround himself with an incredibly powerful energy field. This energy field that was created, according to the scientists' measurements, was equivalent to the energy of a small sun—a mass exceeding one hundred times that of the psychic himself!

As an example, if the psychic weighed 70 kg (154 lbs), he would be capable of creating a concentration of energy around himself equal to that of an energy field emanating from a 7,000 kg (15,400 lbs) sun. Where does he get this energy? Russian scientists theorize that during the psychic's meditative process, he actually concentrates energy that resonates with and draws from the existing "cosmic" bioenergy of the entire galaxy.

The very act of thinking itself is a type of meditation and, as a result, generates a response from the biofield in the form of a corresponding level of energy. This energy generated from the meditative process can then be exponentially amplified through its resonance through the power of existing cosmic bioenergy. As a result, this tremendous energy can effect a change in a situation or even instigate an event that is the focus and intention of the individual's (or individuals') thoughts. Two people, simultaneously meditating, can potentially resonate two times more energy. Likewise an effective meditating group can exponentially generate enough energy to change the aura (biofield) of a large city, resulting in significant shifts as dramatic as the prevention of a natural disaster!

The psychic's meditative state creates around him/her a powerful biofield that is this same energy or force that promotes healing. This in turn restores the health of the patient by opening and reestablishing the unimpeded flow of energy that clears their energetic channels. The new free-flowing "river" of energy can ultimately wash away blockages of stagnant negative energy, resulting in organs and systems functioning more effectively or the shrinkage and complete elimination of tumors, as mentioned.

It is certainly mind-boggling to fathom that by using this psychic technique to tap into this energy source, a healer can direct his hands with intention toward the area of need and actually channel enough energy to eliminate a disease state—even one that has been considered incurable! Furthermore, it has been repeatedly demonstrated that, often in as few as five to ten minutes and at significant physical distance from the patient, I have used this approach and have been able to soften and even decrease the size of many tumors. With each subsequent bioenergy treatment, the patient's energetic status, and ultimately their overall health continues to show improvement.

Seeing the Evidence of Bioenergy on Radiological Scans

Understandably, there are a certain number of people who elect to undergo these bioenergetic treatments with some level of reluctance and doubt. It's usually not the cutting-edge treatment that their physician has been urging them to try. Some feel that it is a last resort or that they have nowhere else to turn. After they do elect to have the treatments and begin to feel remarkably better and/or sense the regression of the tumor through palpation, they are often so dumbfounded and in such a state of disbelief that they respond, "It must be my imagination" "It couldn't be possible," "It was just too easy." This is even more frequently the case with internal tumors, which are neither visible nor palpable from the body's surface. A patient has difficulty believing—and understandably so—that by simply "waving my hand in the air" and holding it in the proximity of the affected areas, I can effect any kind of change. Even after the patient can sense that the internal hardness of the mass has disappeared, and the pain has diminished, it is a real challenge for many people to believe that it was actually correlated with the bioenergy treatments they had been receiving. As a result, the patients will often seek "medical proof" and will request an additional x-ray or CT scan in order to compare the current status of the tumor to its condition prior to the bioenergy treatment. And it is in these post-treatment radiological studies that a peculiar and fascinating phenomenon can occur—one which actually fits with and serves to further validate our understanding of bioenergy. Believe it or not, in many cases, in spite of the fact that the tumor has clearly decreased in size or has disappeared entirely, if the x-ray is taken very soon after the bioenergetic healing session, the x-ray results will appear "unchanged" from the pretreatment studies.

The trace or "ghost" of the tumor—an outline of its shape and position (similar to the results of the Kirlian phantom leaf experiment)—often remains on the latest x-ray or CT scan. I have had some patients whose tumors resolved completely after the bioenergy sessions (this was verifiable through physical examination), yet the x-ray produced very soon after would still "show" that the tumor was there. The concentration of clean, targeted bioenergy replacing the negative energy accumulations would appear as a radiant area on a radiological scan. It appears that the tumor is still present when in fact it is simply the glow of energy concentrated in that region that gives the identical appearance of a mass. In spite of a considerable decrease in or elimination of the tumor and the improved overall condition of the patient's health, many remain skeptical because of their complete reliance on the images seen on radiological studies. In one fascinating instance, the residual energy from a patient's bioenergy treatment just hours earlier was so powerful that it affected the diagnostic machinery and the radiological study had to be postponed. The perplexed technicians couldn't get the equipment to function. Apparently, there was so much energy resonating from the area they were trying to image that their equipment was inoperable! Yet the machine miraculously came back to life and worked flawlessly on the next patient in line!

I am truly grateful to Anush for my profound and extraordinary healing experience. A rush of energy flowed through my body, and then in an instant... months of pain disappeared completely and forever

Claudia Addison, Hollywood, California

CHAPTER 2

The Journey of How I Developed My Ability to Heal

Before actually developing and honing these skills, I too, like any intelligent, rational person, had my doubts about the validity of this technique. Being a physicist myself, not only did I have to see it to believe it, the soundness of the experimental technique was also an important consideration and subject to scrutiny. If someone had predicted twenty years ago that I'd be waving my hands five to 10 centimeters away from a very ill individual's body, rebuilding his or her immune system, and softening and shrinking cancerous tumors in minutes, I would have insisted that it were impossible, labeled it as science fiction and an insult to my intelligence and training! However, an important component of that training is the need to remain open and to have the ability to consider all ideas, regardless of how initially farfetched they may appear. We certainly have enough examples in history of great minds that were initially ridiculed for their hypotheses. If only more of the scientific and medical community could put aside their sometimes one-track ambitions, and protection of their turf, to be open to new possibilities beyond what they were taught in school. As past scholars have told us, it is only through the honest search for knowledge that we realize how little we really know.

Since 1980, I have had the opportunity to demonstrate the power of manipulating the bioenergy field primarily for the benefit of patients and, at the same time, for the validation from the scientific community.

So how does one get to the point of being able to heal using bioenergy? My specific journey as a healer had many twists and turns with my areas of interest and mastery developing and building upon themselves across various mind-body influence modalities, including hypnosis, karate, yoga, folk medicine, homeopathy, shiatsu, and more.

My First Exposure to Hypnosis

My initial interest in the esoteric sciences was actually stimulated by my high school drafting teacher, of all people! At that time, in the early 1960s, *Comrade Martin* was not only an excellent drafting teacher, he also worked with our local police department, serving as a hypnotist and psychic advisor. He was purported to have the ability, through the examination of a photograph, to be able to determine if that individual were alive and their current physical location. If they had passed away, he was able to determine the cause of the death and, in the case of foul play, some details about the murderer.

As his students, we naturally expressed a great deal of interest in this other, far more interesting side of his life. So with a little convincing, he agreed to conduct a lecture and demonstration of some of his abilities. Along with his clairvoyant psychic abilities to sense the locations and condition of individuals, he had the ability to incorporate telepathy into his other area of mastery—hypnosis. This integration of abilities allowed him to give a hypnotic suggestion to another individual without even uttering a word!

Comrade Martin's lecture and demonstration resulted in probably the most memorable and pivotal moment of my entire school experience. He chose as his subject a member of our Armenian basketball team, Armen. Armen was quite the athlete and well-known in our local area as he had also made it to the Armenian National Basketball Team. As a result, he was very popular and well taken care of by many proud fans in our town. He had been bragging to friends that he had been able to get concert tickets that were virtually impossible to obtain. The performance was of a very popular American pianist, Van Cliburn, who a few days prior had won the Tchaikovsky competition in Moscow and had decided to give some performances in several republics of the former USSR. In fact, there were only two Van Cliburn concerts held in the capital of Armenia, Yerevan, so the tickets were extremely sought after and always a complete sellout.

Comrade Martin, "the Hypnotist," had also heard Armen repeatedly discussing the upcoming concert with his fellow students, who were now his audience. So he apparently decided to use that concert as the basis for the suggestions that he would give his subject, Armen.

He began the demonstration by simply asking Armen to stand up and look him straight in his eyes. Armen did as instructed. Then without any lengthy hypnotic induction process or further verbal commands, *Comrade Martin* just stood there and looked at Armen. A moment later, to the utter amazement of his fellow student audience, Armen swiftly reached into his pocket, pulled out his priceless tickets, and tore them in half! He then sat back down in his chair and fell into a sleep. A stunned silence filled the room. There was nothing said that anyone could detect, other than a request for Armen to stand up and look *Comrade Martin in the eye*. How was this possible?

The witnessing of that event was etched in my memory. While that was the beginning of my intrigue with the hypnotic process, 10 years actually passed before I was called to investigate and master the technique myself.

My Interest in Hypnosis Is Rekindled

In 1972, having received my degree in physics from Yerevan State University in Armenia, I had met and married my lovely wife Elada, and became the father of our first daughter, Sona. At that time, we were just starting out and living in a modest one-room apartment that belonged to relatives. The apartment was not only small, but to make matters worse, it was even more cramped—filled with old furniture, miscellaneous knickknacks, and other seemingly worthless junk. In the kitchen, of all places, my relatives kept a large trunk full of old books. As it was consuming valuable kitchen space, I decided to see what treasures it could possibly contain. As I began my investigation, I came across a 1914 Russian translation of a book by a famous American hypnotist, La Mott Sedge. The translation had been printed in St. Petersburg, Russia. The book was designed as a guide for the development of hypnotic technique—written in a very clear, easy to comprehend, and instructive manner. Interestingly, each page ended with a caution statement that said something to the effect of "Do not turn this page until you clearly understand what you have just read." This warning to the reader captured my interest as I read each consecutive page more carefully—making certain to absorb every detail. Little did I know at that time that what I was reading with such concentration and diligence, I would soon have an opportunity to practice and apply it in ways I never would have imagined!

During this period of my life, I had put my career as a physicist on hold for my second passion—working in the Yerevan Drama Theater as a lighting designer. It was simply in my blood, having been raised in and around my father's dynamic work at the theater. I was about 23 years old and supervised five young men and one young lady in the theater's electrical and lighting department. Each morning at work, after studying a section from the hypnosis book in the evenings prior, I decided I would conduct experiments at my workplace. Some of the employees were very interested and were open to the idea of being hypnotized. Others, including one of my young employees, Zorik, had been unwilling for some time. Most of our tests were simple experiments, commanding the individual to lose the ability to move their hands or feet. Generally speaking, once you work with a subject, guide them through a relaxation induction process, and succeed at bringing them into a hypnotic trance, it is possible to leave them with a posthypnotic command, such that in the future you can obtain much more instantaneous results. Zorik, our "scared skeptic," changed his disinterested tune one evening after an event when we were out in the country and away from the theater. Our theater crew had traveled there to assist with a production in that remote area. Zorik had a horrible toothache and was unable to get comfortable and get any sleep in the hotel room. I told him that perhaps I would be able to give him some relief through hypnosis. This time he was a very willing and, fortunately for him, a suggestible participant! Through a slow induction process and hypnotic suggestion about the comfort of his mouth, his toothache subsided, and he was out like a light. From that point forward, he was my biggest fan, and we loved to take a break from our lighting work and entertain ourselves with further experiments to test my developing techniques. I continued to hone my skills, and he turned out to be an excellent and now willing subject. As time permitted, I continually performed subsequent, more sophisticated experiments on him. Even though many of these experiments had no real value in and of themselves, the potential untapped use of hypnosis was very compelling.

Gradually, I began to embellish my experiments from the rudimentary and entertainment oriented to the more sophisticated and life changing. One of the areas that I focused on was the ability to create or elicit from an individual untapped skills and talents. For example, my subjects miraculously developed artistic talents, were able to draw likenesses with the detail and skill of masters, or play a musical instrument they had never before seen or touched!

I was also able to use hypnosis to help subjects find previously lost or concealed objects. Others were induced into having an out-of-body experience where they were able to "travel" to another city and describe details about a place they had never been. In one case, a woman found herself describing a residence in Volgograd (renamed Stalingrad)—known as Hitler's last battle site before his return to Germany near the Volga River, north of the Caspian Sea. She was able to describe this structure in minute detail, including the furnishings, the design and material of the floor of the home, etc. This was more than two thousand miles from Yerevan where the hypnosis was being conducted. Although she initially didn't know where it was she was describing, the location was in fact where her missing belongings had been hidden by a robber!

Individuals who do not ordinarily have psychic abilities to access this latent information can discover these gifts under the influence of hypnosis.

Hypnosis would later become an important component in the development of my healing art.

My Introduction to Yoga

Masters of hypnosis have often recommended the use of the yoga breathing system, known as *pranayama,* to perfect the art of hypnosis. I incorporated this advice and began to study and ultimately experience the impact of these powerful age-old yoga teachings. I started to do yoga in 1972 and practiced everything on my own that I could find in *hatha yoga* books. Then in 1982, I met and received my initial training from a Russian guru from Moscow, Arkady, who was born in India; his father worked in the Soviet embassy in Delhi. He had recently returned to Yerevan where I had the opportunity to study under him. He himself had been taught under a very accomplished and authentic yoga instructor. He learned many secrets, which are not found in any typical books. His guidance greatly contributed to my mastery of yoga.

Through his training, I was able to feel and manipulate each part of my body in a more controlled and deliberate way, while also realizing the health-promoting effects of this practice. While meditating, yoga allowed me to supply my body with life energy to strengthen my human energy field, eliminating negative energy. The yoga breathing exercises, known as pranayamas - i.e., "vital essences" or "breathing forces," as described in medieval European writings - are among the most powerful means known for acquiring and accumulating life energy.

Aside from the physical health-promoting benefits of yoga, there is also its spiritual component. Western thought generally describes yoga as an ancient discipline, practiced by followers of Hinduism and Buddhism, intended to train and transform one's consciousness into a state of perfect spiritual insight and tranquility. Yoga exercises are generally understood to be a part of this discipline, promoting control of both the body and the mind. Historically, the earliest traditions of yoga have tended to encourage a "withdrawal" from the world in order to internalize one's focus and concentration. However, the more contemporary views are consistent with the need to attain spiritual realization without the requirement for such personal isolation. As I have always been one to focus on the practical side of any subject that I study, I have strongly favored this contemporary wisdom, which seeks to positively impact and interact with those around us rather than encourage isolation. While the knowledge of yoga itself—the meditation, finding peace and tranquility, etc.—may be personally rewarding, I felt strongly that the proper way to improve our own "karma" is to use our knowledge and experience to help others. I truly wanted to make a difference in the lives of others, so my belief was that this could be better accomplished through the incorporation of my knowledge of yoga with my other areas of scientific expertise. This ultimately proved to be true and led me further down the path of being able to stimulate and facilitate the healing capacity of another individual in need.

Learning to Heal

There were many steps along my path. One I remember vividly. It was the first time I attempted to really use the various skills and techniques I was piecing together in a way where I instinctively knew I might be able to make an impact. One day, while working at the Yerevan Drama Theater, we received word that one of my workers, Tigran, had been seriously injured in a street fight. No details were provided other than he had been stabbed with a knife and was seriously wounded. The blade had actually penetrated and punctured his lung, and the doctors were not optimistic about his prognosis. This news hit me hard as Tigran was one of my employees, and we were closely connected. He and I had grown up in the same neighborhood and recently become quite close. As the doctors were monitoring his condition, even after the third day of hospital care, he was showing no improvement. Becoming even more concerned at their obvious helplessness in the situation, I felt that I needed to at least attempt to influence the situation using the healing philosophies I had been immersed in for over 10 years.

Through my study and practice, I believed that using the appropriate visualizations, meditations, and intention, I could perhaps promote healing in him from a distance. That evening, upon returning from work, I decided to make every effort to connect with him. I first found solitude in a separate room and asked my family for some time to concentrate. After spending some time meditating and centering myself, I imagined Tigran in front of me. With my intention and focus on him, I entered into a deep meditation. I imagined his wound as best as I could. Using the limited information I had received about its details, I also used whatever psychic impressions about it that materialized during this state of openness. Once I had the wound clearly in my mind, I began to visualize "cleaning" it and filling it with healing energy. After spending some time in that "cleaning state," next I began visualizing the closing and sealing of the wound. Finally, I imagined that I was able to influence the growth of new lung tissue into the damaged area. All of this I did in this state of deep meditation and concentration and with a feeling of connectedness and caring for the young man. I truly felt that I had been connected with him and hoped that there would soon be good news.

The next morning, upon arriving at the theater, I was immediately greeted with the buzzing among the theater workers that Tigran's condition changed! In fact, it had taken a dramatic turn for the better! The doctors were dumbfounded and couldn't account for the change as nothing had really been done from their perspectives. They were considering it a miracle. He had regained consciousness, and it soon became apparent that he would make a complete recovery.

Tigran's recovery was the first of many more opportunities that followed where I began to develop and put this "new" healing art to the test.

Meeting a Mentor in the Study of Bioenergy Healing

While we were still living in Armenia, Gia Mepisashvily, a young, well-known spiritual teacher from Tbilisi, the capital of the Soviet Republic of Georgia, was conducting a lecture circuit across the Soviet Union. He was well-known for the esoteric school he had founded, known as *Transcendental Medicine.* Gia's travels brought him through Armenia, so I was very fortunate to have the opportunity to attend his talks. The philosophy of transcendentalism itself, upon which his teachings were based, was certainly not new at that time. The German philosopher and metaphysician Immanuel Kant, American author and philosopher Henry David Thoreau, and English authors, such as Carlyle and Wordsworth had certainly discussed and widely debated this philosophy in the early to mid-1800s. However, it was the relationship to medicine and healing that Gia had incorporated that was intriguing.

Transcendentalism emphasizes individualism and self-reliance. Included in its doctrine is the belief that God's eminence is found in both man and nature and that an individual's intuition is the highest source of knowledge. Transcendentalism maintains that *"man has ideas that come, not so much through the five senses, or the powers of reasoning, but are the result of direct revelation from the God, his immediate inspiration, or his eminent presence in the spiritual world."* It asserts that *"man has something more than the body of flesh: he has a spiritual body with senses to perceive what is true, and right, and beautiful . . . and a natural love of these things."*

Having the opportunity to hear Gia Mepisashvily's lecture was a privilege, and it provided so many new insights and ideas I had never considered, yet now presented, suddenly seemed so obvious and crystal clear. Then being able to meet him, and ultimately work so closely with him, led to some of the most dramatic changes in my life. His approach to healing, through transcendental medicine, seemed to be the link I had been seeking. It was the "glue" that brought all of my previous scientific study, experience, and insights into a relationship or model that gelled.

Over the course of several months, under Gia's training, I was finally able to meld all of my years of theoretical and philosophical knowledge and pursuits and really put them into useful practice. Gia worked closely with me to fine-tune my meditative and healing practices. With his guidance, I was able to perfect the technique of opening the nine "chakras" in our body. The chakras are considered to be energetic centers for the respective area of the body that they control. They can be thought of as the "power plant that supplies a region with electricity." Using this technique, I am able to harness energy from the universal energy field and distribute it via the chakra energy centers to that part of the body that the respective chakra regulates.

Many of the beliefs and practices developed by Gia had been indoctrinated by his late grandfather, who had been a student of George Ivanovich Gurdjiev, a once-prominent WWII-era intellectual, mystic, author, and dancer. Gurdjiev, born in Alexandropol, Armenia, spent much of his time at Fontainebleau, near Paris, where he founded an intellectual group known as the *Institute for the Harmonic Development of Man*. Gia's grandfather traveled with Gurdjiev and was able to study and ultimately adopt many of his ideas. After Gurdjiev's death, Gia's grandfather continued to develop and expand upon Gurdjiev's philosophies, which interested Gia, and later became the springboard for his beliefs and teachings.

Gia's school, and his personal guidance, led me to develop my powerful meditation abilities and healing techniques more fully. I slowly perfected the meditations and visualizations that allowed me to manipulate the human energy field. That period of development was the most pivotal part of my healing training and study, thus far, and probably more significant than all my years of study to date. Of course, all of the experiences built the foundation and led me down a path to ultimately mastering and imparting these healing techniques.

Over the next few years, I developed the capacity to both feel and actually control the bioenergy contained in an individual's energy field. As a result, I was able to diagnose patients, both in person, and at some distance, using photographs, telepathically, or through speaking over the telephone. I learned to, in a sense, "clean" the individual organs and to consequently heal a broad range of illnesses. Within one year, I was healing hundreds of people suffering from a myriad of illnesses including digestive, neurological, endocrine, and other disorders. Over time, I was eventually able to treat various forms of cancer with ever-increasing success. Every case was truly a unique experience, each one revealing new avenues of exploration for bioenergy therapy. It was an exciting time in my life. The impact I was able to have on the lives of many people was rewarding, and I never ceased to be amazed by the actual simplicity of it all.

My wife, Debbi, was diagnosed with stage 4 multiple myeloma and her doctors could only offer painful treatments and a quick death sentence. After her Bioenergy Therapy sessions with Anush It felt like she had been given her life back. She felt so good that we wondered if the doctors had misdiagnosed her.

Tim Guisinger, Camarillo, California

CHAPTER 3

My Wife Elada's Breast Cancer

It's OK to Heal Everything—Except Cancer

As mentioned, from the time I started to use my healing abilities, I treated many conditions by a technique that essentially consisted of meditatively opening the patient's obstructed energy channels, restoring the flow of that energy, and finally rebalancing the patient's chakras or energy "distribution centers" of the body. I will outline many of these cases in the next section. However, approaching the treatment of cancer was a very different challenge. Yes, it was considered incurable, so you might say, "What do you have to lose by trying?" However, that is where the problem lay. There was a strong and frightening caution to stay away from cancer because of the healer's apparent vulnerability to contract the cancer.

Coming Face-to-Face with Cancer

As I alluded to in the introduction to this book, shortly after the birth of our daughter, Sona, my wife, Elada, discovered a tumor in her left breast. At that time, I was still progressing in my abilities as a healer. Her doctor, a well-respected family friend at the nearby Armenian X-ray

Institute, recommended surgery. Even so, after much consideration and discussion, we decided against the procedure.

We came to that conclusion in late December as the winter vacation was commencing at the local schools, which coincided with the ramp-up of the winter theatrical productions. Elada, who was an actress, was invited by the Armenian State Concert Organization to star as the "Snow Maiden" in a winter children's production. The five daily shows were to be performed in two adjacent halls of the huge movie theater with only 30 minutes break between each show. That means 10 shows daily, five in each hall. Elada, thrilled and honored by the invitation, accepted the challenge enthusiastically. Unbeknownst to me until some years later, part of her acceptance of the demanding role stemmed from her concern over our family's future financial situation in the event that her health were ever compromised. So she immersed herself into the rigorous daily schedule of rehearsals and performances, running from one hall to the other. The production actually created a festive, uplifting atmosphere with the many dancers, comedians, musicians, clowns, jugglers, acrobats, and other entertainers participating. It was a full day for Elada with the shows running from 10 a.m. to 8 p.m. She arrived home exhausted yet exhilarated. The nonstop schedule and fun-filled atmosphere seemed to distract her from her fearful ruminations about the tumor and allowed her to redirect her thoughts on the excitement and accolades she was receiving from her performances and the spirit of the holidays. It was ultimately the best possible medicine for her emotional and physical state. After the series of shows concluded, and she stopped to catch her breath, she realized that her depression had actually lifted. Having the starring role in the production and all of the associated demands ended up being health-promoting! Her positive outlook and upbeat attitude stayed with her over the next several years even with the tumor.

For the nine years following the initial discovery of the mass, I had been significantly building my knowledge and expertise as a student of healing. Still the practice of bioenergy was thought to be ineffective in the elimination of cancer and, more importantly, dangerous to the healer who attempted it. Even so, I decided to go ahead and try my skills at Elada's tumor despite my fears and the many cautions from my mentor and others in the healing field.

My first two bioenergetic treatment processes consisted of two courses of ten bioenergy treatment sessions. I focused on directing the powerful, cleansing bioenergy into the tumor area to "clean" the negative energy and attempt to flush out the tumor. These first two attempts yielded no apparent change. We weren't going to give up so easily! I continued with a third course of 10 treatment sessions, which proved to be much more encouraging! The tumor appeared to soften slightly. So we pressed on with subsequent treatments. The results were gradual but clearly evident. There was a continuing decrease in the size of the tumor and a change in its texture. Finally, after a total of approximately three courses of 10 bioenergy sessions, the initially resilient tumor eventually completely disappeared. It was astounding!

I have since continued to observe that same phenomenon that often longstanding tumors, regardless of their size, tend to be more stubborn in their responsiveness to bioenergy healing. I hypothesize that those tumors, because of their longevity, have had more time to actually change the surrounding physical tissue, making them more difficult, yet not impossible, to affect. Our persistence in that case was rewarded. Today, Elada is as healthy and beautiful as ever; and we realized that cancer is not invincible.

I'm sure you can imagine how overjoyed we were that Elada was clear of the cancer. It was also a significant milestone in the use of bioenergy healing techniques especially with the previous warnings to avoid any such attempt. My breakthrough was communicated to others studying the mastery of these techniques. Even so, many still felt that it was a big hurdle to even consider taking on!

I believe that the most critical element that resulted in our success was not just the use of the powerful meditation technique that I had learned from Gia and now perfected but, more important, the fact that I was able to clean, strengthen, and guard my chakras while coming face-to-face with her cancer in this battle. The focus on protecting my chakras during this process gave me the ability to defend myself against any potentially dangerous consequences that may result from going after the cancer head-on. I had essentially learned to "wall myself off" while performing the healing, and because I am still able to do this, cancer is no longer invincible.

Protecting Myself

Even though I had developed this technique of shielding myself during the bioenergetic treatment process, it is not always foolproof. I later found myself shocked and in panic when I encountered a situation where I actually contracted the very disease of my patient.

A paralysis patient, Gurgen, was having regular ongoing bioenergy treatments with remarkable success and, in his enthusiasm, was spreading the word about his progress. He was a very credible man—a well-known physicist and honorary member of the Armenian Academy of Sciences. He asked if I would be willing to consult with one of his guests, Alpik, who was visiting during one of our sessions. Alpik had been diagnosed with lymphatic cancer in the Kremlin Hospital. He too was a physicist and the director of the House of Scientists, the establishment where all fundamental physics research was conducted.

I took a look at Alpik and, upon assessing his condition, was actually quite alarmed. His entire lymphatic system appeared damaged with large tumor growing under his left arm. Alpik reported that he had already been to Moscow for a diagnosis and prognosis. The physicians had apparently given him a life expectancy of approximately three months.

Even with his very conservative physics background, he was open to any new ideas about how he could heal. Particularly with his friend and colleague Gurgen's improved paralysis symptoms, he was very grateful for the opportunity to also receive a bioenergy treatment session. I immediately directed the bioenergy into Alpik's tumor in the hopes that we might see an immediate change of some kind. The result astonished everyone—including me! After that single session, the tumor had diminished noticeably, and his feeling of well-being and energy improved. We were all overjoyed by the rapid result, and I left the two gentlemen and returned home. I expected to perhaps conduct some follow-up sessions with Alpik to see how the tumor remission would progress.

Shockingly, the very next day, I discovered a tumor—almost the mirror image of his—under my left arm, in the very same location! At this point in my training as a healer, I felt very inadequate, as I had no idea what to do in this predicament. So I looked within to my yoga background. Having studied yoga for many years, I kept myself in excellent physical condition. So I tried for several days to muster up my energy to remove this tumor. Nevertheless, the size and hardness of the tumor remained constant. I meditated and ruminated about what I could do to resolve this horrible consequence of my good intentions. Suddenly, I recalled an old story I had heard of a similar nature. There had been a yogi who had a tumor and had been unsuccessful in removing it for some time. Finally, he decided to try another approach. He enlisted the assistance of a friend, who happened to be a surgeon. He asked him to interpret the "cryptic" x-ray so that the yogi could recognize the location of the tumor image. The surgeon showed him the blurred area that represented his tumor. The yogi then asked his friend, the surgeon, to make a small incision to mark the area where the tumor resided. The surgeon complied although he had no idea where all of this was going. The yogi left the surgeon's office and headed straight for a nearby forest. He located a tree that was of a size that he could wrap his arms around in an embrace. At the same level of the tree, where his wound was, he removed the bark from that section. Then he aligned his wound with that bare part of the tree and pressed his wound against it while embracing the tree fully with his arms. He began to meditate, visualizing the dissolving of the tumor and the reabsorbing of it by the tree. He continued to do so for several days. Finally, he instinctively felt as though it had resolved, so he returned to the hospital for another glimpse. To the great surprise of the doctors, there was absolutely no evidence of the tumor.

I had heard this beautiful story some years ago as an early student and hadn't thought much of it at the time. However, it's interesting how you can pull something out of the deep recesses of your memory banks on just the occasion where it may be potentially lifesaving! I decided to experiment with this same concept with a slight modification. In my case, I chose not to make an actual incision under my arm at the tumor's location. I entered the forest and searched for a living tree branch of about two to three centimeters in diameter. I bent the branch and then made a diametrical cut through it. I then took the cut "naked" section of the branch and pressed it against the area under my arm where the tumor was. My hope was to use the idea of transferring the negative energy accumulated within the tumor into this severed tree limb with the belief that the energy would then continue through the tree itself, down though the roots, and into the soil. With the cut branch in place, pressing on this new, freshly acquired tumor, I started my biomeditation. I focused all of my thoughts, intentions, and concentration on removing and transferring the negative energy accumulated in the tumor through the tree branch all the way into the soil. I continued this meditation for several days. Finally, to my great relief, the tumor began to show signs of decreasing in size. Little by little, it continued to diminish until it had completely disappeared.

What an overwhelming feeling of relief! I had been able to eliminate my tumor through the corroboration of an old teaching, using a similar natural way of healing. Even with the happy ending, Alpik's case has become a powerful reminder for me. Although I was miraculously able to eliminate Elada's cancer without any consequences, this was a strong reminder to approach all of my cases more cautiously, paying close attention to the cleaning and protecting of my energetic system, both during and after my healing sessions.

Alpik, whose tumor appeared on my body and who had been told he had three months to live, was alive eight years later when I lost contact with him as I left for the United States.

Over time, my practice became more focused on the treatment of cancer patients who have limited useful mainstream medical options—unlike the more successful treatments that might be available for those with other chronic diseases. Also, as a physicist-experimenter, working with and trying to uncover more of the unknown, I always wanted to try to undertake what was considered impossible. So cancer treatment was the ultimate and became my specialty.

Healing Cases

In my years of treating illnesses using bioenergy, both as an old-world healer in Armenia and later in the heart of Los Angeles in my Santa Monica Boulevard office, I have experienced a number of startling successes that have amazed even me. I have encountered cases of "terminally ill" patients under the care of prominent, highly regarded, even famous, doctors, who have realized absolutely no improvement. Yet, on the contrary, with some aspect of bioenergy therapy, these same patients, who are considered hopeless in their challenge with their disease, have recovered. Sometimes my treatment was as simple as just opening their energy channels or applying pressure to the appropriate points on the body, which coincide with that organ system or body part. In other instances, inspiring words sometimes reinforced by hypnosis can create the belief in the patient that he can be healed. This in itself can bring about a full recovery.

After Years of Chronic Kidney Pain, Albert Now "Treats Himself" Using Bioenergy!

One of my colleagues, Rafik, took me aside one day at work to talk about his brother Albert, who was in a great deal of pain. Albert was a long-distance truck driver and had been experiencing ongoing flare-ups of severe lower back pain, a known occupational hazard in his line of work. His doctors wanted to admit him into the hospital for testing. Albert was avoiding that recommendation and continued to suffer through the bouts of excruciating pain. As it turned out, Albert lived just a block from my house and was having one of those chronic episodes, so I was able to respond quickly that day to see what help I could provide. When Rafik and I entered Albert's room, he was flat on his back, in obviously debilitating pain, and moaning. He had a number of people at his side—his wife, his mother, and a few friends— all well-meaning, but helpless to give him any relief. I approached the bed where he lay so uncomfortably. I gently asked him if he was able to find any position where he was able to feel any freedom from pain. With an obvious strain heard in his voice, he uttered "no" as he was struggling to turn over on his stomach—a seemingly desperate effort to find any contortion to quell the pain.

Once he was settled again, I scanned his back and the rest of his body. It was clear that there was heaviness in the kidney region. I began to focus a bioenergy treatment on the lower back and directed bioenergy to that stagnated kidney area to "clean it." I continued this for only several minutes before Albert began to show signs of response. His moaning began to subside, and soon he became very still and quiet. I continued the treatment, and a few moments later, he actually began to sit up in the bed. He had a look of relief through his exhaustion as though he had just completed a big undertaking or crossed the finish line of a race. He sat in his bed peacefully while I continued the kidney treatment. I began to sense a significant improvement in the region and asked him if he'd like to try to stand up. He looked at me tentatively as though he were really pushing his luck. It had been a long time since he had been able to sit comfortably. He didn't want to revert to another episode of writhing pain. I assured him that he would be fine. He cautiously inched his body to the edge of the bed and then, very deliberately, hung one leg at a time over the edge. Then he gingerly placed one foot on the ground, slowly followed by the other. He kept testing his pain, amazed that he felt none.

Finally, he actually put all of his weight on his feet, stood up, steadied himself, and then took several steps on his own. We were both somewhat astonished. He felt no residual pain in his back and kidney area. His friends and family were breathing sighs of relief and amazed that the cries of pain had faded, and a sense of peacefulness had returned to the house.

As it was getting late, Rafik suggested that I head home and that he would accompany me. After reaching my house, he thanked me profusely and then eagerly headed back to check on Albert's progress. By the time Rafik returned, the calm in the room had transformed into a party atmosphere with Albert drinking vodka, and talking up a storm with his friends in celebration of his healing!

Albert, both grateful and curious, contacted me sometime later to learn more about how I had managed to make such an immediate and dramatic difference in his condition, where his physicians had been unsuccessful with surgery as their only recourse.

Albert and I arranged to meet again to talk further about the treatment, and as a result, he decided that this was the type of healing to which he wanted to subscribe. So Albert became one of my most diligent and enthusiastic students of bioenergy therapy. I gave him an introduction to my techniques, which he began to practice religiously. With his ongoing self-treatments, he was able to maintain overall vibrant health and continued to be permanently free of the once debilitating pain.

A Three-Year-Old "Destined" for Institutionalization

Early in my healing practice, I naturally approached my work with children very cautiously. Even though we know that bioenergy therapy contributes to restoring health and an ideal state of balance in a person, there is always that aspect of dealing with a young, precious child that makes us all handle them with extra care and concern.

The first child who was ever brought to me for treatment was Armen, my friend Arto's three-year-old-son. Little Armen had recently suffered an asthma attack that left him in a much-weakened state. As you know, asthma is becoming more and more prevalent in many children today with so many environmental pollutants. The subsequent bombardment of a myriad of medications, many of which are immunosuppressive, can cause a host of other complications. Since the asthma attack, Armen now quickly became fatigued and winded with minimal exertion and couldn't play and romp around with the other youngsters as he had been accustomed to. Arto and his wife were against the use of many prescription drugs and wanted their son to have a way to heal more naturally using his immune system. So we began a course of treatments to work with and help him along. After Armen's first two healing sessions, his parents called excitedly to report that his energy was returning, so much so that they were thrilled to report that they couldn't get him to sit by himself quietly anymore! He had actually "turned the apartment upside-down!"

It was clear that he was responding very favorably to the treatments, so we continued with regularly scheduled sessions. Not only did he regain his energy and was once again a typical three-year-old handful to his parents, but his asthmatic coughing also completely subsided. Armen returned to being a completely healthy three-year-old, full of life and curiosity.

One of the other wonderful aspects specific to treating young children is that through their openness, they can somehow sense the healing process going on around and within them. That in itself is very comforting to them, such that even when they are in pain, they become immediately calm and attentive. Remarkably it is a complete contrast to the stress that is frequently observed in children in a typical sterile medical office treatment setting.

Anna Has Unexplained Neurological Disorders

My mother-in-law had asked me to do "just a little" spring construction project on her property. I can now look back and see that isolated request as the beginning of the path that ultimately led me to encounter an infant in need of healing -- an infant who was eventually to become my goddaughter. What a blessing!

In order to transport the construction materials to the site on my mother-in-law's property, I needed a small cart of some sort. I called a friend, Yury, who had some connections in the village. He always knew who to call for just about anything.

We spoke about my need for some sort of materials transportation, and he said he'd look into it and get back to me. He called several days later and said he'd arranged to pick up Elada and me and drive us from Yerevan to the small village of Egward, some 15 miles away. There he said we could get a cart from the village veterinarian, who also happened to be his childhood friend. As you can see, nothing was all that simple in those days! Yury picked us up as promised, and we drove out to his friend Artsrun's farm. Artsrun was very accommodating, and he showed me an assortment of carts and welcomed me to choose the one that would best suit my needs. I selected one and then Artsrun loaded it in his truck and offered to transport it to my construction site in the next day or two. Then instead of heading back immediately to Yerevan, Artsrun insisted that Elada, Yury, and I come up to his home to meet his wife, Inessa, and his family. Inessa had unexpectedly prepared an elaborate spread. We enjoyed the wonderful food, our newest friends, and their sweet young daughters. The youngest, just 16 months old, was obviously not well. During our entire visit, she lay virtually lifeless in her bed—definitely not what would be expected of a child of that typically curious and energetic tender age. Just before we were preparing to leave, my wife pulled me aside and asked me to look at the child to see if anything could be done for her.

Earlier, while they were in the kitchen, Inessa had confided in my wife that there was a problem with her youngest child, Anna. Inessa said that she was born with what was vaguely described as "deficiencies" that were unexplained by their medical experts. Anna responded only passively to those around her and lacked control of her body. Inessa said that doctors had advised them to institutionalize Anna. The child was unable to sit or even raise her head and had a great difficulty in swallowing food. She had minimal neuromuscular control or stability.

I went over to Anna's bed and attempted to sense the energy state in her very still body. What was immediately apparent was that starting with her neck and continuing throughout her entire spinal column, there was a very limited or constricted flow of energy. The first logical step then was to open up this channel so that the flow of energy would be unimpeded and restored. Using bioenergy, I directed it through her neck and spinal column, flushing out the channel and filling it with restorative bioenergy. After about 15 minutes of this process, I rechecked her energetically, and she was completely different. At that moment, I didn't detect any more obstructions there or elsewhere. So I returned to our gracious hosts, and we said our good-byes and set out for home in Yury's car. During our drive home, Elada whispered reprimands to me about not devoting more time and attention to little Anna, who was obviously in such great need. I explained to her how quickly the blockages in the spine yielded, and within just 15 minutes, there was a free movement of energy. I had removed whatever that obstacle was impeding her energy flow. I felt in my heart that there was going to be some kind of a change.

The next morning I was back working on my construction project. In the midst of it, a large truck approached, and I realized it was our new friend, Artsrun. He was already bringing me the cart that we had chosen to borrow the day before from his farm. He pulled up and hopped out of the truck with a big smile on his face. He single-handedly hoisted the cart out of the truck and wheeled it over to me. We exchanged our greetings, and I thanked him again for his and Inessa's generous hospitality. He said to me, "You know"—turning and pointing to a nearby tree—"that branch would be ideal for hanging a lamb." I had no idea what lamb he was talking about. He proceeded to tell me that on Sunday, after church, he would be making a religious sacrifice of a lamb and felt that this house would be "the best place to have the party." This house? I was certainly in favor of the festivities but was perplexed as to why he had chosen here—15 miles away from his home in Egward! Artsrun went on to explain, in a soft and moved voice, that after the relatively brief healing session that I performed on their daughter, Anna was already completely different. After we left, they began their usual bedtime preparations for Anna. Inessa asked Artsrun to hold their little daughter while she arranged Anna's bed. As Artsrun held her, instead of her usual deadweight and listlessness, their daughter began to crane her neck and began to wriggle around in her father's arms as if to try and free herself. Shocked at the unexpected movements, he nearly dropped her.

"Inessa, Inessa," he shouted. "Look at Anna! She is trying to get away from me . . . to get down to the floor." Inessa didn't react immediately. She thought he must be joking, but then quickly realized what was really going on. Inessa stopped in her tracks with her eyes glued on Anna. Anna began to move her body, as if she wanted to stand or sit on her own, struggling to pull away from her father's secure clutch. She had never done anything like this before. Artsrun and Inessa were in complete awe. They were so struck with amazement by her demonstration that they didn't want to put her to bed so quickly!

That night they prayed that this had not been an isolated event, but that it would be the beginning of further signs of an awakening in Anna. They looked forward to the next morning, hopeful that Anna's newfound signs of determination would continue.

Sure enough, by the next morning, her activity level increased even further, and she started to exhibit movement in her hands and legs. She even started to raise her head from her customary weak and drooped position. Artsrun wanted to stay home from work as he and Inessa were mesmerized by their child's truly miraculous progress. Inessa couldn't take her eyes off Anna all day, and after work, Artsrun rushed over to my home with the report of what had transpired. After witnessing this they knew that it was important to continue the bioenergy treatments on a regular basis. Over the course of approximately two months, all of little Anna's neuromuscular function and coordination was restored to that of a normal child her age.

We never knew exactly what had caused the obstruction in her energy that created such a severe handicap. However, she has now grown up to be a completely healthy young girl with no limitations, and she gave me the extra special honor of becoming her godfather.

A Breast Tumor That Disappeared in 10 Minutes!

One typical evening, as I was walking home from work on my usual route, a car pulled up alongside me and stopped for no apparent reason. I was initially startled when the door opened and then recognized the smiling face of one of my old karate classmates, Yervand. It had been years since we had seen each other. We began chatting about old times, and I realized he had a passenger with him. He suggested that I too jump in the car, and he could give us both rides home. His young friend, Anahid, was about thirty and sat very quietly, seemingly preoccupied and melancholy. Yervand began to explain that just a few days ago her doctors had discovered a tumor in her breast. Suspicious of cancer, they had recommended surgery to remove it immediately. Anahid was naturally distraught over the sudden news. Her worries were compounded because of her seven-year-old son, who, for the last few years, had grown up without his father.

Coincidentally, Yervand had suggested to her that she consider seeing me and had told her about my success with healing. They had planned to try to locate me, and then, out of the blue, there I was on the road walking home! "Talk about divine intervention!"

As the car neared my house, I asked if they would like to come in and have some tea and visit. They were delighted with the idea.

Elada was there. She greeted our unexpected guests and then managed to throw together some refreshments on short notice. We all ended up sitting around for what ended up being hours drinking tea, snacking, and chatting about everything.

We totally lost track of the time and then realized it was after midnight by the time we finally decided to look at the clock. We hadn't even gotten around to the topic of Anahid's tumor. Yervand pointed out that it was probably time that they headed home. I said, "Well, before you leave, let me at least examine Anahid." Anahid said that the tumor was large enough for her to feel on her own, which was what had brought her to the doctor's office earlier that week. My examinations, being energetic in nature, can be performed with a patient fully clothed. Anahid stood in front of me at the door while I sensed her energetic flow. It was true that there was a detectable stagnation of the energy flow, which correlated with the tumor location in her breast. It felt that the area of the tumor was about the size of a small nut. While still standing there at my front door, I conducted a very brief bioenergy healing session of just five minutes. I then asked her to examine herself for any noticeable changes. She examined herself quickly, as it had been easy for she and her physicians to find the lump even through her blouse. This time, she seemed to be having a little more difficulty locating it, and she looked at me with a bit of a startled look. Then she continued her assessment by partially unbuttoning her blouse and putting her hand directly on the skin surface for a more thorough examination. She appeared momentarily confused. She then mumbled something about checking herself more closely at home and swiftly started heading out the door. Yervand followed quickly after her to start the car.

The next day, Yervand called to apologize for their abrupt departure. He said that he had spoken to her that morning and that Anahid said the tumor was gone! She couldn't find any sign of it. He said that she was in a state of obvious shock over the entire experience and was completely lost for words. Later she called and explained as well—now elated! Initially after the session, and not finding any sign of the tumor, she was in such a perplexed state that she didn't know what to say and apologized for not expressing any words of gratitude at that moment. I told her no apology was necessary. This was a common response of patients when the results are so instantaneous. I was so glad we had such an immediate response.

From time to time, I'd run into her on the street, as her home was near my workplace. She continued to report happily that she was feeling wonderful and that the tumor had disappeared.

Hypnosis and Bioenergy Therapy—A Powerful Combination

Another former classmate, Nune, called me at the theater, distressed. We had been very close in school, like family. Over the years, our families continued to gather for the holidays, our birthdays, the birthdays of our children, etc.; however, not many of them were cognizant of my healing practice as I rarely spoke of it. They were, however, aware that I practiced yoga, karate, and a little bit of hypnosis.

Nune called, upset about her six-year-old niece, Sirooshik. Sirooshik's parents had returned from Cuba a few months prior, after working there for two years. Just two weeks after their return, the little girl started to experience unexplained pain in her arms, starting from her shoulders and continuing to the tips of her fingers. The pain worsened with each passing day. Along with the discomfort, her fingers began curling to such a degree that she would bind her hands into tight fists that wouldn't release. Each attempt to loosen and open up Sirooshik's fingers would result in pain-filled cries from the little girl.

Sirooshik's parents had since taken her from doctor to prominent doctor in Moscow, Leningrad (now Petersburg), Kiev, Riga, etc., and none of them seemed to be able to determine the cause of these peculiar neuromuscular symptoms.

The pain had now progressed to her legs, and she was unable to walk.

Nune explained, "We have been feeling completely lost with absolutely nowhere to turn. Sirooshik had to start school, and she can't use her hands. Then today, out of the blue, one of my friends was talking about someone he knew that had extrasensory abilities and could perform veritable miracles . . . and with a little luck, we could probably track him down. As soon as he gave me your name, I started to laugh. I had no idea that it was you and that you were a renowned healer! He couldn't get over the fact that I actually knew you! I was so relieved when I found out it was *my Anush*!"

We set up an appointment for me to work with Sirooshik right away. Nune explained to me that to encourage and comfort Sirooshik, she had told her that they were going to visit no ordinary doctor, but a magician who makes you better without bad-tasting medicine or painful shots. So Nune asked if I could try to be mystical yet serious, if possible!

I promised to transform myself into a magician for our sessions! My office with its six-meter ceilings and ornate furniture and curtains made it an appropriate backdrop for the "magic" about to take place. When

Nune arrived with little Sirooshik, her eyes were understandably filled with fear. And it was no wonder after how physically uncomfortable she had been. I welcomed them, and Nune, along with her sister Luci, sat down on the couch quietly. I sat in an armchair and gently asked Sirooshik to come sit with me. Luci began to recount their endless search for medical assistance. While we were talking, I evaluated Sirooshik's condition. I found the energy in her legs, hands, and spine to be very heavy and not flowing well at all. I asked her to look me in the eyes and then tell me if anything hurt. Little Sirooshik bravely said in a pain-filled voice that her hands and legs were hurting very badly. I explained to her that it was because her hands and legs are filled with *a type of dirt* that I could magically make disappear, and the pain would vanish along with it. Sirooshik seemed to accept that; and I began cleaning her spine, legs, and then the back of her hands. After a few minutes, I detected an improvement in the energy flow, so I asked her how she felt. She hesitated in surprise and then in her little voice said, "It doesn't hurt right now." I continued on with a second level of "cleaning."

I then asked if she could try and open her fists for me. She tried, but said it still hurt.

I asked her to look in my eyes and gently explained that there was still a little more dirt trapped in her hands, and as soon as I removed that, she would be much better. I promised that after the final cleaning, the pain would disappear. I focused on thoroughly cleaning her closed fists by directing bioenergy through the energy channels. Finally, I was ready to ask her again; however, this time I looked into her eyes and simultaneously gave her a hypnotic suggestion: "Sirooshik, you will now be able to open your fists without being afraid, and they will feel fine." Cautiously, she opened one hand very slowly and then the other—as if to test the waters. Not experiencing any pain, she looked relieved and finally opened them completely. A big smile appeared on her sweet little face. I asked her to open and close her hands several times, which she did. I then reinforced the hypnotic suggestion by saying that from this point forward, she would always be able to open and close her hands without any pain.

From across the room on the couch, her mother Luci had tears in her eyes. Then her response surprised me. She grabbed her daughter by the hand and hurriedly exited the room without uttering a word. Her sister, my friend Nune, raced out behind her, saying that she would be sure to call me the following day.

Nune called apologetically, explaining that the experience was so overwhelming that they could not manage to thank me. (As you may have noticed, this is common.) She said that Sirooshik was feeling so better and that her parents were still in awe over what had transpired. The distress they had endured for six long months had disappeared in a matter of minutes. We jointly decided to bring Sirooshik back for several follow-up "magic shows" to ensure that we kept the pain at bay permanently.

In this particular case, Sirooshik's healing was achieved by a combination of bioenergy therapy and hypnosis. The hypnosis was helpful because when a patient has experienced pain for an extended period, they sometimes become "programmed" to feel the pain long after the actual cause of the pain had been eliminated. So hypnosis is a helpful adjunct to pain management as the body is still in the process of completely readjusting itself.

A Four-Year-Old Boy with a Malignant Neck Tumor

I was still working as a lighting artist in the theater. Late one morning, one of our accountants, Lusine, pulled me aside in an obvious state of grief and asked me if I could find a little time to examine her friend Anahid's four-year-old son, Laert, who had accompanied her that day. Young Laert had a very prominent and energetically heavy tumor on the left side of his neck. In my estimation, the tumor was clearly not benign.

About six months prior, when the tumor had appeared, Anahid and her husband had taken Laert immediately to the local oncology institute. After some analysis, the doctors diagnosed it as a rapidly progressing cancerous tumor and offered to start chemotherapy immediately. Anahid and her husband were unwilling to accept this course of action right away and opted to take him to Moscow for another evaluation. The diagnosis was unfortunately reconfirmed, and chemotherapy was again prescribed. Anahid categorically refused as two years earlier, her five-year-old niece had been diagnosed with a similar condition and given similar treatment plan. Her niece had been subjected to this standard prescribed treatment—beginning with chemotherapy, followed by radiation. The little child suffered tremendously and, sadly, survived for only a year.

As a result, Laert's mother was disillusioned and would not support any mention of chemotherapy. She asked the doctor proposing the chemotherapy if anybody with Laert's condition had lived any length or quality of life following the treatment. The doctor explained that chemotherapy was not a cure, but just a mechanism to prolong life. She said she needed a doctor who believed in trying to find a way to actually restore real life to her son. She returned home looking for other answers.

By the time she brought Laert to me, her son's condition was deteriorating at a rapid pace. His appetite was poor and he was becoming easily fatigued. It was nearly noon that day, and the rehearsal was already commencing on stage, so I was pressed for time. I grabbed the first free room, so we could get some privacy and examine Laert. The tumor, which was positioned on the neck, seemed to have roots spreading into the surrounding tissue. The young boy was surprisingly calm and accommodating, so I began his first bioenergy treatment session that day at the theater. Although the tumor was very visible and solid, after the first treatment, it already seemed to be loosening up from the "root system" to which it was attached. It actually was dislodging itself and had become more mobile, shifting a bit to the left and right and up and down. Anahid was intimately familiar with the every contour and feature of that mass, having witnessed its growth and change over the last six months. Anahid felt the tumor and immediately recognized the weakening of its hold. Tears of joy and hope came to her eyes. She suddenly felt that God had heard her prayers and had led her to an answer.

It was now time that I had to join the rehearsal in progress, so we agreed that Anahid would bring Laert to my home daily for a while to keep the momentum going and work toward getting the upper hand on this tumor.

Over the course of several days of repeated treatments, there was a definite change in Laert. He started to regain the natural energetic spirit, curiosity, and appetite of any healthy child his age. It was especially gratifying to witness this unfolding in my home. He and my daughter were close in age, and he'd want to engage in games with her and was especially excited about riding her tricycle. After all of that activity, he'd announce to me that he was hungry, and that was music to everyone's ears as we were more than happy to cater to that request.

After about the fifth or sixth session, not only was the tumor releasing itself from its tangled encroachment of the boy's neck but it also was softening and shrinking! He was responding beautifully to the bioenergy treatments. Anahid wanted to continue the sessions until there were absolutely no signs of this malignant tumor.

Typically I conduct about ten sessions and then take a break and allow the body's immune system to do some of its own work. Through the treatments, it is now regaining its strength and becoming reenergized. Over the course of that year, we would follow this cycle of resuming treatments and taking subsequent breaks, until finally, the now five-year-old Laert was cancer-free!

As many Armenians were headed for America, Anahid was considering it as well and had a visa prepared. She had relatives in Montebello, California, ready to receive her family. Her husband, a master jeweler in our town, was not particularly eager to make the move, as he had a flourishing business. However, when my family began to apply for emigration, Anahid decided to take Laert and her family to California as they felt concerned that Laert may need further healing at some point in his life. Fortunately, that was not the case. Laert, like many young children, responded beautifully and permanently to the treatment. I did stay in touch with the family and watched Laert grow into a fine healthy young man. At 21 years of age, I remember him as one of the tallest and most athletic among boys his age. What an incredible feeling to see that young boy, who was able to ride a tricycle for the first time in my home, enjoying his teen years without any limitations! Thankfully, he remained well; however, one of Laert's young friends ended up needing care and learned of my bioenergy healing from Anahid. I will discuss this case in a later chapter.

Seda Recovered from Terminal Lymphatic Cancer—Her Doctor Drops His Glasses!

Bioenergy therapy influences the progression of cancer cells at any level or stage of development and even in circumstances where there are other parallel treatments being performed. However, ideally if a patient receives bioenergy therapy *before* a mainstream treatment is sought, I believe that the patient will have about an 85 percent chance of a full recovery. Bioenergy treatment is often shown to be more challenging after the patient has undergone a course of chemotherapy and definitely after any radiotherapy. These patients not only need to be treated for the cancer, but they now also need "cleaning" to rebound from the adverse effects of these medical treatments.

One such case involved a patient, Seda, the assistant to the mayor in the Armenian resort city of Dilijan. She lived near the area known as Sevan Lake, where many of the residents knew her and the fact that she had cancer. When I met Seda, she had just returned from Moscow where she had finished a second course of intense chemotherapy at the Kremlin Hospital. With a diagnosis of cancer of the lymph nodes, this lovely young woman, exhausted from these taxing, ineffective treatments, appeared as though her fire were about to extinguish. She told me that shortly after the first course of chemotherapy, her condition actually worsened. The cancer spread, so the doctors' solution was to order a more intense course of chemotherapy. It's not uncommon for cancer to progress in these situations, as, sadly, these treatments actually undermine the body's ability to heal because they take such a toll on the immune system and other supporting systems. By this time, her desire for food was nonexistent. Her breathing was labored, having to stop and rest every few minutes when she walked or exerted herself even minimally.

She had come to see me, very skeptical and reluctantly, and only as a result of the insistence of one of her relatives. He too was employed with the city government in the auto inspection area. He had encouraged Seda to seek my help as he had previously recovered from a serious kidney condition through our sessions using bioenergy.

Initially Seda and I had only four days to work together as she was scheduled to return to her business in Dilijan to meet with a delegation arriving from Moscow. I hoped that even in the course of four days of treatment, there would be some tangible improvement—and there was. Seda's trip back to Dilijan was much easier than anticipated, and she arrived home feeling remarkably better, considering her postchemotherapy condition. There was such a positive turnaround and newfound sense of well-being versus the struggle to function she had been previously enduring, that she became quite optimistic and decided on her own to return for additional sessions the following week. This time her outlook and belief in the possibilities of bioenergy had come full circle, as she was now talking about being cured one day! We moved forward with another full course of bioenergy treatments.

Each subsequent day she reported feeling better and better. Her appetite and energy level continued to increase. Her strength was returning both physically and aerobically. Seda would venture out for walks without the prior exhaustion she had experienced after just the first few steps. However, most important, she had dismissed her doctor's pessimistic prognosis and instead truly believed that she could be cured—and that, from such a skeptical and practical woman, was a real testimony to how comparatively different she really felt. We worked together for 10 days in this second course of treatment before she returned home to Dilijan to begin working again.

Many of the residents of Dilijan, and her colleagues at the mayor's office, were staying abreast of her condition and were also hopeful about her strides to recovery. Seda shared with me, with great enthusiasm, the details of her first day back at the office.

Her office was on the fifth floor of the regional committee building. With no elevator in this building, she frequently had to use the staircase several times a day. Upon her return, many of her colleagues stood tentatively, watching from the landings between the first and fifth floors. All of them remembered well how challenging it was for Seda to negotiate those stairs. They watched with joy and amazement as she made her way up to the fifth floor without the assistance. As she easily reached the last step on the fifth floor, a huge round of applause and cheers burst out from her elated colleagues. Seda equated it to feeling like an Olympic athlete crossing the finish line for the gold!

Over the course of that year, Seda completed several subsequent courses of bioenergy treatments at my home. That summer, she graciously invited me and my family to come for a visit and stay in one of the resorts in Dilijan. We had a wonderful trip, and we made our way around the area, meeting many of the townspeople. Seda was practically a household word as she was loved by the people of Dilijan, for her very kind, philanthropic, and humanitarian contributions in her community.

We also had the pleasure of meeting her wonderful husband and three-year-old daughter. Her husband, *Vardan*, and I spent some time together while Elada got to know Seda. Vardan shared with me in confidence that after the second course of chemotherapy in Moscow, the doctor pulled him into the office and told him that he needed to know that Seda wouldn't live for more than two months—and that he'd better make the appropriate preparations. Sitting in that doctor's office, he was in such emotional turmoil, crying uncontrollably, feeling helpless and so wanting to save his wife somehow. He tried to collect himself to protect Seda from seeing his grief; however, she caught him exiting the doctor's office and knew instinctively what was in store.

Now, instead, they expressed gratitude for the great blessing of her recovery.

Seda eventually returned to Moscow to check in with the doctor who had given her the earlier terminal two-month prognosis. As she entered his office, standing before him, he did a double take in disbelief, dropped his eyeglasses, and was momentarily speechless. Not only was she alive, but also obviously well and in good spirits. Once he reoriented himself, he welcomed her visit, expressed his enthusiasm for her rebounding from the illness, and then suggested a repeat of the diagnostic tests to reevaluate the status of her lymphatic cancer. Once the results were returned, he and the entire Kremlin medical staff were astonished. There was no further trace of the cancer—not in the lymph nodes, organs, bones, or anywhere that they examined. Seda and her husband were ecstatic!

Zara's Lupus: A story as told by her father, Rudik Karabekov, renowned nuclear physicist (paraphrased and translated from his book, Conversations)

Searching for a New Healing Modality for Our Daughter

There have been psychic and other healers known to perform miracles. I myself was an eyewitness to one of these kind of miracles, attributed to a healer, Anushavan Manukyan. He did something unbelievable, something extraordinary—a dream for all medical doctors, scientists, and institutions:

Zara, at the age of 18, acquired a serious illness called lupus. The doctors were perplexed for some time, initially unable to diagnose this obviously very serious condition. Her lungs, heart, and body were full of fluid. She was feverish and had an irregular heartbeat. It had been two years since she contracted the disease, and the only treatment was still to use prednisone, a steroid that helps with some of the symptoms. There are a number of side effects to this treatment, including masculinization, osteoporosis, and consequently, joint deformities from the calcium depletion. The first sign was a deterioration of the hip joints making it difficult to walk. If this continues patients become invalid and bedridden.

The prognosis from "the best doctors—rheumatologists and orthopedists of Yerevan and Moscow" was extremely discouraging. They had little to offer and decided that her best course of action was for her to replace her hip joints with prostheses. Even if that route were taken, it was in no way a guarantee that further problems as a result of the disease and its treatment would not occur.

We felt there had to be some way to help our daughter, so we continued to pass the word around, asking for any ideas from friends, family, acquaintances—anyone! We thought if we communicated it widely enough, perhaps someone might have an answer for us. In fact, that turned out to be the case when I shared my daughter's plight with my colleague, Artick Arvanov, a scientist, who worked at the Institute of Physics. The answer from Artick fit perfectly like a lock and key. Artick listened intently and then recounted the following miraculous story of the healing of a tumor over his son's kidney that prompted us to also seek Anush's help.

We Learned of Tomas's Discharge from the Army with an Adrenal Tumor and Anush's Miraculous Healing!

Artick's son *Tomas* had recently been discharged from the army. While in the Russian military, the more senior recruits, who had been in service about two years, beat him up so badly that he ended up in the hospital with a damaged kidney. An operation was required to repair the injury. While repairing the damage, the surgeons discovered an adrenal tumor on his right kidney. As a result, he was subsequently discharged and returned home. With each subsequent day his condition deteriorated, and his parents took him to the Kremlin Hospital in Moscow. The tests done at the hospital confirmed that the kidney did, in fact, have a tumor the size of a fist. An operation was to be scheduled but was repeatedly postponed for two months because of the sheer number of patients crowding the hospital. In the interim, they returned to Armenia where the son's condition worsened, and his father decided to look for help from extrasensory healers.

Artick remembered that he had a friend living in Kiev, who had a great deal of experience and knowledge about healers. When he attempted to telephone him, he discovered that his phone, as well as some others in that area of Kiev, were out of order because of flooding from a heavy rainstorm the previous day. Artick resourcefully decided to try the telephone system of a friend who lived outside of the city. When he arrived to use that phone, his friend was having a large gathering in his home. Artick made his way through the party guests to telephone Kiev. Thankfully, he did connect to his friend in Kiev, only to find out after his steadfastness that his friend couldn't be of any help. However, with the poor connection to Kiev, Artick found himself yelling into the phone in order to be understood, such that the entire party could hear the conversation clearly. One of the guests, overhearing, interrupted Artick and said perhaps he could help. He gave Artick the name of *Anushavan Manukyan*. Artick appreciatively rushed home immediately with the lead. That same day, he and his wife contacted Anush and brought Tomas to him for an evaluation. After a consultation, he told them that their son had a tumor on his right kidney the size of a fist. The parents were in shock at the preciseness of his diagnosis, as they had not yet shared the details of their son's condition with Anush.

So impressed, they decided to immediately proceed with bioenergy healing sessions and put Tomas's surgery on hold. After only six sessions, Anush sensed that the tumor had shrunk to about the size of a walnut. In their excitement and newfound hope, the family flew to Moscow to visit the original doctor at the Kremlin Hospital who had diagnosed their son's tumor. The doctor agreed to repeat the tests to assess the current situation. To the doctor's surprise, the radiological studies showed that the tumor had in fact shrunk to the size of a walnut! Artick and his wife were in complete amazement as the family hurriedly returned to Armenia to continue their son's healing sessions. After another four sessions, the tumor had completely disappeared. The family returned to Moscow once again to get a final round of tests done at the Kremlin Hospital. Unfortunately, the doctors were not enthusiastic about repeatedly performing tests on a patient who was not receiving treatment there yet was improving. It ultimately required bribery (a sadly common occurrence) to get the doctors to agree to retest their son. The tests fortunately gave them the peace of mind they needed. The tumor was completely gone! The Kremlin Hospital physicians were unable to explain the comparative studies of the old and new films. There was no sign of the tumor on the new films. The doctor's "conclusion" was: misdiagnosis by an erroneous x-ray. A mistaken test, or, in this case a battery of tests, did not explain his weakness and inability to work prior to the healing sessions. Not only did the tumor disappear, but Tomas's vitality, health, and well-being were restored. No longer able to serve in the army, he enrolled in the university physics and mathematics program that year and later graduated with honors.

Zara Meets Anush

This section continues from the perspective of Rudik Karabekov, Zara's Father

When my colleague, Artick, recounted that it was his strong and literally loud intentions resounding through the phone and over the party noise that resulted in his prayers being answered for another healing path. He was blessed with the discovery of Anush's healing gifts and the resulting miracles. This prompted me to go home immediately and speak with my wife and daughter Zara. I then made an appointment to see Anush that evening.

Upon meeting Zara, Anush admitted that he was not familiar with lupus but was very willing to do anything he could to help her.

Upon examining Zara, Anush's scan of her body immediately revealed a generally very heavy biofield. He reported that all of her bones and limbs seemed to be affected by this disorder. She demonstrated a great deal of difficulty walking, and even though she was a very young woman, she had already been using a cane for quite some time. As he continued to evaluate her, he detected what he termed as "heaviness in her biofield" in areas including the spine, feet, heart, and liver. Even so his gut said that he could make a positive impact on her condition energetically.

With each daily session of him doing his part, he also taught her a combination of yoga and karate exercises that she could begin to learn to strengthen her muscles and restore her strength and energy. After the first few daily sessions, there was already a significant change. Her bones had actually stopped aching, and she was already able to walk without the cane! At one point, she excitedly told Anush that she had attended a friend's birthday party and danced! Zara continued to improve, and her ability to return to many of her usual activities and pace of life steadily progressed.

The Doctors Said, "We've Never Seen Lupus-Related Osteoporosis Reverse Itself!"

Over the course of six months, Anush performed 75 bioenergy sessions! It is remarkable that there is no hands-on contact necessary. At the end of this extrasensory energy marathon of sessions, Zara's walking improved dramatically, and she felt better than she had in years! One of her doctors, Sergay Agababov, who was one of the best rheumatologists in the Soviet Union, wanted to re-x-ray Zara's hip. He was preparing to attend the International Rheumatology Conference, taking place in the famous Soviet Union port and health resort of Sochi, on the Black Sea. Many of the world-renowned specialists in this field would be on hand. Dr. Agababov wanted to present Zara's case and get the opinionsof his peers and discuss future treatment protocols. He planned to show a comparison of Zara's initial x-rays with her most recent x-rays, which reflected the effects of the bioenergy sessions. Ten days after his presentation, Dr. Agababov called my wife and me ecstatically and stated that something "unbelievable . . . a miracle" had occurred! It was the first time in medical history that osteoporotic bones had undergone a restorative process. That is, no one at the conference had ever heard of a case where the osteoporosis associated with lupus had not only been stopped in its tracks but was beginning to reverse itself! This was basically considered impossible, as there was a direct, medically well-documented correlation between the continued use of steroids for their anti-inflammatory and immune system-suppressing effects, and the recognized side effect of bone loss. After Dr. Agababov's enthusiastic and miraculous report, I shared with the doctor that it was no accident, as she had been receiving bioenergy treatments from Anush for the last six months.

After a pregnant pause, Dr. Agababov seemed to admit that it was not likely anything in his standard treatment program that resulted in Zara's remission. He then gave an encouraging word and agreed that the bioenergy treatments were obviously valuable and that the medical field apparently still had a lot to learn and discover about the human body.

Anushavan Manukyan opened up our minds and even those of the medical community to demonstrate that even the "impossible" was possible!

- Shared by Rudik Karabekov, Father of Zara

Zara and Family . . . Years Later
- from the author's perspective

Zara frequently visited us often with her parents, and our friendship continued until I moved my family to the United States. I recall, during their last visit to my home, Rudik's heartfelt expression of gratitude. He said I had "returned his lost daughter to him, and it was as if she had been reborn." Zara demonstrated her newfound agility as she swooped down effortlessly to pick up her father's keys as they dropped to the floor—without even bending her knees!

Her gracious father credits the bioenergy work we did that allowed her to be able to reach her true potential. Several years ago, I invited her to visit our family in Los Angeles. I was honored to find out that her father, Rudik, had written the aforementioned book,_*Conversations*, including a lengthy chapter about lupus and my ability to eliminate the debilitating symptoms of the disease.

Zara went on to graduate from college, marry, and give birth to a beautiful, healthy daughter, Alisa. Zara was then accepted to graduate school in Yerevan, Armenia. Soon after, the United States started building an elementary particle accelerator and began seeking a very high-level nuclear scientist to direct the program. Her father, Rudik, was approached and, upon his acceptance, was granted U.S. citizenship for himself and his family. As a result, Zara transferred to Eastern Virginia Medical School in Norfolk, where she still resides with her husband and her daughter. She chose to study biomedicine with a specialty in immunology and cancer. Her intent was to learn as much as possible about lupus. On May 18, 2001, she presented her thesis and received her PhD in biomedical science.

When Anush raised his hand, I could actually feel the gentle energy pass through my gall bladder, eliminating the years of intense radiating pain and restoring my digestive capabilities. Astonished, I became a student of his methods and a sponsor of his workshops so that others could discover their self healing abilities using the amazing bioenergy field that is all around us!

Irene Vincent, Laguna Niguel, California

Chapter 4

Bioenergy Begins to Capture the Attention of the Soviet Government and Scientists

As the word of the healing work that Gia and I were performing was being circulated, the Soviet government as well as local scientists and government officials were among those who were hearing of our successes and, as a result, began to develop an interest in our techniques and the mechanism behind them.

"Visitors" from Moscow

On one of my mentoring visits with Gia in Tbilisi, we had an interesting turn of events, resulting in one of our more memorable visits. Gia's girlfriend Theresa's father was a well-known sculptor. Because of their visibility in the art community, her parents knew many influential people in and around Tbilisi as well as Moscow. Theresa's mother, Yadviga, called Gia telling him that two "visitors from Moscow" had arrived that "wanted to meet the renowned Gia Mepisashvily." She had apparently talked to them at length about Gia's abilities, and they wanted to make the trip to Tbilisi to investigate. Gia accepted their request to meet and asked if I'd join him. Yadviga and her husband, the sculptor, lived in an upscale apartment in one of the more desirable parts of

Tbilisi. When we arrived, the guests were already there having tea. We received a warm greeting from Yadviga, and she presented us to the two visitors: one a major in the KGB, the prestigious Russian security and intelligence agency, and the other a director of the *Zero Institute*. The two gentlemen explained that there was a widespread Soviet effort to investigate, study, and document claims of various extrasensory occurrences and stories of gifted individuals throughout the USSR. This was the first Gia or I had heard of this. The director went on to explain that additional branches of the *Zero Institute* were being established across the Soviet Union.

The two gentlemen went on to describe their individual investigative responsibilities under this program and told us it was they who had "validated" the abilities of the now-famous Juna Davidashvily. As a result of their discovery, she was subsequently brought to Moscow to demonstrate her energetic abilities. Juna Davidashvily was known to everyone in Tbilisi as she had been born and raised there. She led a typical life growing up in an Arabian family. She was a good student, graduated from high school, and then began working as a waitress in a local café. There appeared to be nothing extraordinary about her until it became known that she could open champagne bottles without touching the cork. The restaurant patronage began to soar. She would reportedly bring her hand close to the neck of the bottle, approximately one or two inches from the cork, then suddenly fling her hand upward, and the cork would explode and shoot out from the bottle.

Would-be customers, hearing about her great "trick," would come in droves to buy champagne, just to see her perform.

The café management had no idea about the special skills of this seemingly ordinary waitress. They were more than happy to begin stocking cases and cases of champagne as the little café turned into a gold mine. Eventually Juna began to channel her psychic and energetic abilities far beyond opening champagne bottles, so her "performances" at the café and the owner's associated windfall wound down. Juna discovered that she could psychically heal a variety of diseases and was asked to come to Moscow to demonstrate. Word spread and she became so well known to the point that her "patients" included many highly respected senior Russian officials. At one time, it was rumored that she was treating Leonid Brezhnev, President of the USSR at that time.

During this visit with the Moscow officials, the KGB major told us that he had been notified about medical "miracles" that Gia had performed. He wanted to meet with Gia and hear about his abilities and the supporting theories behind them. As we sat with the two gentlemen, Gia explained the basic concepts of the human energy field and how bioenergy could be used both as a diagnostic tool and to promote healing. The two officials listened with great attention and interest. Then, seeming to understand the mechanisms, they then asked me to specifically demonstrate my ability to diagnose a medical condition. I began by scanning one of the gentlemen and assessed and reported to him his currently "heavy," or potentially diseased areas. I also identified other areas which had been a problem in the past. He was highly impressed with the accuracy of the assessment. During these demonstrations, the second official from the *Zero Institute*, although he had remained relatively silent in the background, was observing the results very closely. We explained that with this type of sensing, distance is not a factor. So they skeptically asked for me to evaluate his wife who was miles away, and of course, neither Gia nor I had ever met her! Once again, the results spoke for themselves. In astonishment, he exclaimed very enthusiastically that what we were doing was "fantastic!" He himself had been involved with the study and practice of parapsychology all his life and had reached the point of being able to levitate an object, such as a package of cigarettes using the palm of his hand as a guide. Yet he was in awe—aspiring to achieve such a level of sensitivity as our medical diagnostic capabilities demonstrated.

Validation by the Scientific Community in Yerevan, Armenia

Just as interest came from the prestigious *Zero Institute*, similarly, the local medical community was equally drawn to our work. A large group of scientists, doctors, and students of the Medical Institute located in Yerevan, Armenia, convened in an auditorium. They had requested my presence to demonstrate and explain my ability to shrink tumors. A woman with a scientifically validated breast tumor had been asked to be the first subject for the demonstration. Before beginning my "treatment," her palpable breast tumor was measured. I then proceeded to focus on dislodging the blocked energy that had manifested itself as this tumor. The quiet auditorium of onlookers remained quiet as I conducted a session, which lasted only about ten minutes. At that point, I felt that there was a substantial change in the energy flow as a result of the treatment. I gave the OK to Dr. Amatuny, who was overseeing the demonstration, to have a look.

He immediately measured the tumor again and then, in a very official tone, announced the new measurement, which was about half of the original size. The medical and scientific community present was astounded as the tumor, in a matter of just minutes, was verified to have decreased by approximately 50 percent. This demonstration was representative of the rapid response that many tumors show to the induction of bioenergy. A patient's response varies according to many factors, including how long the tumor has been present, such that in some cases, which I will discuss further, some tumors completely disappear in a single bioenergy treatment session, whereas others are more resilient and require repeated treatment.

Requests from the Medical Community to Diagnose Hospital Patients

Soon after those compelling results in their auditorium, I received an invitation from Professor Amatuny, the Dean of the Yerevan Medical Institute, to come to the hospital clinic and give a demonstration there of my diagnostic abilities. Professor Amatuny was not only a highly educated man and excellent physician, but he was also known for his open mind and forward thinking. He took a special interest in alternative medical approaches and the occult sciences. The professor was quite familiar with my mentor Gia's's healing techniques. He had heard of me only in the context of being one of Gia's more gifted students. Naturally I accepted this next invitation to accompany him to the hospital!

A room had been set aside at the clinic for the patients to be screened. Upon our arrival, the clinic was jammed with both interested doctors and students trying to make their way into the evaluation room. Twenty patients, who had already been admitted to the hospital and had been diagnosed using traditional methods, were brought into the room for my assessment. The doctors and students assembled themselves in the room. Once he had the attention of the clinicians, the professor began to make several announcements. He introduced me to the group and explained that because I was not a medical doctor, he made it clear that he did not expect me to have the ability to label the illnesses by their proper medical names. He asked that I simply identify any ailing organs or bodily functions. A hush came over the room as I approached the first of the 20 or so patients. With each patient I quickly passed my hands across and just above the surface of his or her entire body. As I detected certain vibrations or areas of specific "heaviness," I expressed my findings to the professor, identifying particular regions or organs. Based on some of the whispers and mutterings coming from the onlookers, many of whom were already familiar with these patients' maladies, I realized that my level of accuracy had captured their attention. The professor was taking copious notes as I moved from one patient to the next. Finally, after completing my "diagnosis" of the last patient, the doctor reviewed his notes and made some calculations. He then announced to the audience that I had exceeded his expectations. Remarkably I had been able to properly identify nearly 95 percent of the known ailments with some patients having multiple conditions. The audience and patients applauded with enthusiasm and amazement, now seeing the overwhelming evidence of bioenergy's potential application as a corroborative diagnostic method.

A short time after this medical assessment session, Professor Amatuny contacted me and asked if I would be willing to collaborate with him in the future, especially on some of the more difficult new diagnostic cases. This seemed quite exciting and fulfilling to step in and help this physician who was one of the few rare individuals sincerely interested in learning all aspects of alternative medicine and the esoteric sciences if he could. And at the same time this collaboration would broaden the overall credibility of alternative methods of diagnosis and healing. He and I developed a very effective working relationship that grew from mutual respect into a strong connection and meaningful friendship.

Psychic Surgery in the Philippines Also Uses Bioenergy Techniques

As I spent time with Professor Amatuny, I learned that his enthusiastic interest had arisen as the result of his brother's very personal and compelling experiences. His brother was a professor and top surgeon at Moscow's Kremlin Hospital. A highly intelligent and educated man, Dr. Amatuny, like his brother, was very open and inquisitive about learning about healing techniques, folk medicine, Eastern thought, yoga, and more. One of his desires was to visit the Philippines and see a surgical procedure performed by one of their famous traditional Filipino healers. These unusual operations were performed without a knife. During a visit by Dr. Amatuny to the Soviet embassy in the Philippine

Islands, this dream became a reality. Coincidentally one of the Embassy officials told him that he himself had a small benign tumor, which he wanted to have removed using this psychic surgical method. So he invited Dr. Amatuny, in his role as a reputable surgeon from Moscow, to satisfy his curiosity and accompany him to witness it firsthand. Dr. Amatuny readily agreed to join the official, and an appointment was made for the "surgery."

He and the Embassy official entered the healer's "office." The healer, about forty-five years of age, introduced himself as *Adan* and talked to them briefly about what would transpire before directing them to the "operating room." With minimal assistance, the Embassy official was positioned on the operating table as *Adan* made his preparations to begin. Suddenly, he seemed to become distracted and stopped in his tracks. He said he was not going to be able to perform this operation on the Embassy official at this time as there was a more pressing matter. Instead he turned to and wanted to speak to Dr. Amatuny, who was seated in the corner, attentively watching. He spoke frankly to the Soviet surgeon, and told him that he had a lump on his chest and that he wished to remove it right away. Dr. Amatuny was startled that this announcement came out of nowhere. Apparently, the healer began to pick up psychically on Dr. Amatuny's medical condition, which was not what they had on their agenda that afternoon! The healer was correct. Dr. Amatuny did in fact have a large mass of what he believed was a fat deposit on his chest. It was not particularly bothersome, certainly not visible to the healer under his shirt, but nevertheless there!

Dr. Amatuny gratefully gave his permission to attempt to remove the mass. Positioning Dr. Amatuny in a reclining position on the "operating table," the healer wrapped the fingers of his right hand around the deposit. With a few deft movements of his hand, it was as though he cut the skin with a laser beam and lifted the deposit gently away from the body, holding it up for Dr. Amatuny to see. Dr. Amatuny was utterly amazed. His entire traditional medical training had been shattered in that moment while, simultaneously, his instincts to look more closely at alternative modalities and the occult sciences were forever etched in stone. The Filipino healer performed this surgical procedure quickly and cleanly without instruments, blood, or discomfort—and, more important, without a lasting scar that can lead to a potential disruption to his bioenergy flow.

Dr. Amatuny immediately reported back to his medical institution the turn of events and miraculous procedure. The Soviet government, in turn, immediately responded and instructed Dr. Amatuny to stay in the Philippines and master this remarkable, instrument-free surgical technique. After several months of study with the local healer, he felt competent and returned to Moscow to share his newly developed skills. The word got out about his return, and he was soon invited to the Academy of Science for a demonstration. The event took place in the largest of auditoriums, filled with doctors and medical students. The crowd waited attentively to witness how it could be possible to perform surgery using one's hands without wielding any sharp instruments!

The demonstration was set on a small platform with the patient prepared for the procedure on the surgical table. The surgeon, using his now well-practiced-as-second-nature "invisible scalpel," guided his hands across the patient's skin, using bioenergy in an attempt to sever the skin. He paused, bewildered, as for some reason the skin remained intact. He tried again repeatedly. There was no response. After several minutes of no response, he took a break and retreated from the surgical table to engage with the audience. He answered a few general, polite questions about these procedures, and then planned to attempt the technique again; however, the disillusioned audience began to exit.

At first, it was difficult to understand how his proven abilities to do psychic surgery in the Philippines were not repeatable in this Moscow setting. Professor Amatuny recounted this dilemma with me about his brother Dr. Amatuny's failure. After some further questioning, I concluded that the obvious catch was that the surgery was successful in the Philippines because of Dr. Amatuny's ability to tune into and access the other powerful energies and validating belief systems present in that environment. His energy was being reinforced, replenished, and amplified by the other more-experienced healers around him. Conversely, in Moscow, and in an audience of many skeptics, he had no such energy reserve or access.

Collaborating with and Inspiring Professor Amatuny to Add Bioenergy to His Medicine Bag

During one of my collaborative diagnostic sessions with Professor Amatuny, *Mihran*, a slightly overweight man in his mid-40s, presented himself in the professor's office complaining of a pain on the left side of his abdomen. The professor spent a good deal of time examining him. Compressions around the stomach region were triggering substantial pain. Unsure of the source of the pain, Professor Amatuny gently told *Mihran* that it may be necessary for him to remain in the hospital for a period of about five days for detailed testing in order to get to the root of the problem. Then the professor paused for a moment and turned to me. After all, he had brought me on to collaborate and provide insight into his diagnosis and evaluation of patients. He asked me if I could take a look as well and see if I could identify anything specific from my perspective.

I approached *Mihran*, who was noticeably uncomfortable, so I tried to be very quick and conducted a rapid energy scan of his body. He, of course, was at a loss as to what I could possibly be doing! I then took Professor Amatuny aside to consult. In my estimation, I believed the source of the "heaviness" and associated pain was originating from the rectal area.

The professor, pensive for a few moments, asked me if I would go ahead and attempt a bioenergy therapy session in the area that I thought was problematic. If there were any change, even a subtle one, we'd know at least we were headed in the right direction, which might at least hone or minimize any necessary diagnostic testing.

After just a few minutes of focusing energy to the rectal region, *Mihran* reported that the pain had noticeably backed off. He moved around a bit and compressed the area of his abdomen again, looking for remnants of the pain of which there were none! He was incredibly surprised—the professor even more so. Professor Amatuny too began a series of gentle compressions around *Mihran*'s abdomen inquiring as to whether he had any pain. Still he reported none. Finally, the professor asked him to lie flat on a massage table where he could be a bit more aggressive, using firm yet careful compressions. Once again, he remained pain free, and the professor could do nothing but willingly release him.

Soon after that collaboration, Professor Amatuny decided that he was really driven to understand bioenergy to the fullest extent. He began regularly attending the lectures of my mentor, Gia, and as a result, immersed himself into the study and eventual adoption of his philosophies and healing modalities. Professor Amatuny went on to eventually become the first medical doctor in the USSR to diagnose patients using his hands alone! Professor Amatuny, to his credit, was one of the rare medical doctors who remained open to the many alternative treatments and approaches, even while continuing to practice modern medicine in the high-tech world!

Armenian Government Officials Seek Healing

In 1982, I was paid a visit by two unidentified men. They very mysteriously relayed a message that someone had told them about my healing abilities. They continued to speak in vague terms about a "very important person," who suffered from chronic and severe back pain. The pain had been present for so long that the individual was unable to recall the last time when he was able to stand erect comfortably.

They went on to say that this patient was open to any measures or conditions to which I was agreeable, if I would be willing to intervene and guarantee a positive outcome. Although we well know there are no guarantees in life, I felt it was certainly worth attempting. His identity was still a mystery, and I was intrigued by the fact that he had sent these emissaries to pave the way. We promptly set up an appointment, and the mysterious man and his wife arrived at the arranged time.

The gentleman and his wife were very pleasant. He looked very physically fit for someone who was battling so much pain. I was to later learn that this unidentified man was the head of Transportation for the Interior Department of Armenia, a high-ranking position. His name was Colonel Edward Shahinyan. His wife, Yuna, also had a respectable position in the OVIR (department of emigration service). She was in charge of the department that issued visas for travel to the socialist countries like Poland, Romania, Bulgaria, and Czechoslovakia.

My wife and I welcomed the Shahinyans and invited them to have some refreshments while I was assessing the colonel. One of the remarkable outgrowths of gaining so much experience in the mastery of bioenergy treatment is that just about anything can be going on around, and you still have the ability to perform the treatments. There needn't be silence in the room or any particular ambiance. It's just a matter of being able to make the meditative connection, which with practice can become instantaneous and possible in virtually any seemingly distractive surroundings.

I spent some time thoroughly diagnosing him and discovered that he had disc and nerve damage. He corroborated the assessment, sharing that his back problems developed as a result of a combination of years of heavy lifting and then sitting for long periods of time in confined positions, etc. We discussed a treatment program and my recommendation to begin a regular course of bioenergy sessions along with some shiatsu, and that these could be conducted either at his residence or at mine.

We decided to alternate between both locations. What an unexpected pleasure to visit his beautifully furnished apartment in the center of town that made it feel as though I were spending an afternoon at a museum. There was a large white concert piano in one corner of the room. Above it hung a very large original oil painting by the famous Russian artist, Ivazovsky. Ivazovsky, as you may know, painted mostly seascapes and shipwrecks—perhaps the most famous being *The Ninth Tidal Wave*. However, the colonel also had others in his collection, including a rare atypical Ivazovsky painting, portraying a shepherd sitting under a tree, playing the flute, surrounded by his flock of sheep.

I continued to treat Colonel Shahinyan over the course of several months as he finally began to experience relief from what seemed to have been an eternity of back pain. He was very grateful, and my wife, Elada, and I developed a wonderful and ongoing friendship with him and his wife in the process.

During one of our then-social visits, the colonel asked if I'd consider a favor. His next-door neighbor, Tital (another government official in charge of the city's Department of Motor Vehicles), was very proud of his "perfect" health; however, the colonel knew that something wasn't right with Tital. I agreed wholeheartedly to meet the gentleman if he too was agreeable.

One phone call from the colonel, and a few moments later the gentleman arrived. On the surface he was a very energetic man of about 50, and athletically built. We had our introductions and the colonel explained my diagnostic-sensing abilities and the relief he had personally realized for his chronic pain. The colonel's longstanding struggle was well known to those close to him, including Tital. The proud head of the Motor Vehicles Department agreed to have his very healthy constitution evaluated, not expecting any abnormalities to be discovered. He made himself comfortable, chatting with the colonel and his wife while I silently conducted my diagnosis. It was apparent that in fact, physically, he was very strong. At the same time, there was heaviness in his kidneys. I concluded my assessment, complimented him on the obvious care he took of himself, then mentioned the fact that there could be an issue of kidney stones brewing, and suggested that perhaps he have his doctor perform the appropriate diagnostic tests. I would be more than happy to stand corrected, however. I felt that there was something there worth investigating.

He found my comments to be amusing and without foundation as he hadn't the slightest inclination of a problem in any organ system, including his kidneys. He had always been very healthy and physically in perfect condition. We went on to visit for the remainder of the evening and then parted ways. The next day, I received a frantic call from the colonel's wife, Yuna. Obviously in a state of concern, she rambled on about Tital having had a sudden and serious attack in his kidneys. The ambulance was called. However, the emergency medical technicians attended to him for some time and were unable to provide him any relief from the pain. Finally, they saw no other choice than to transport him to the hospital, where hopefully more could be done. Tital was very uncomfortable and, at that moment, moaned about urgently needing to empty his bladder. In that uncomfortable process of urinating, he passed a sizable kidney stone. At that moment, his pain promptly subsided. Fortunately, he was then able to be stabilized and recuperated nicely.

Interestingly the news of my sensing his problem developing just days before in this man, who was well known to everyone as the picture of health, spread rapidly throughout his office. By the following day, the entire Department of the Interior was buzzing about it. The news then continued to spread throughout the community, further substantiating and spawning more interest in my "unique diagnostic and healing abilities."

Using a Bioenergy assessment, Anush alerted me to conditions and imbalances in my body with remarkable precision that were previously asymptomatic and unknown to me. As a result, I was able to take measures to investigate/research and address these health concerns before they could compromise my well being.

Jacob Hoban, Ph.D., Natural Healing Center, Encino, California

CHAPTER 5

Our Plans to Move to America: My Healing Art Paves the Way

Even with all of the attention and unmatched satisfaction that came from helping so many people in need in Armenia and the surrounding area, it was my family's dream, as shared by many, to move to America.

An Appreciative Father Brings Us Closer to America!

I had developed a friendship with a gentleman, Leonid, who was a former classmate of my cousin, Tatool. We had gotten to know each other attending karate classes together. Leonid had gone on to work for the chief of the OVIR (emigration service). Our friendship had become quite close in the early years when one of his family members became one of my first healing success stories. So when I began to prepare to come to the United States, Leonid was immensely helpful with my emigration paperwork. At that time, it was virtually impossible to leave the country without paying a bribe of five to six thousand rubles (average monthly salary was 120 rubles) for each family member leaving the country. This unethical requirement was considered commonplace and expected. Many people did this willingly because, even after selling all their belongings, any additional cash could not be transferred abroad. Each person who was granted permission to leave the USSR destined for the United States was only able to take $45 with him.

So it was one of those unfortunate bridges that the average person would have to cross in the hopes of a better life for their family!

After applying for my family's visa to leave the USSR, I became a frequent visitor to the OVIR to get questions answered and to ensure I was following the process appropriately, navigating through the emigration maze. My friend Leonid's knowledge of the ropes was a godsend. He'd frequently invite me into his office to discuss our emigration and make sure everything was in order. On one of the visits, he happened to introduce me to a young man about 35 years of age by the name of Arkadi. Arkadi was a representative of the KGB within the OVIR where Leonid worked. It became evident that Arkadi's permission was a requirement for the approval of each and every passport or visa. Through our seemingly peripheral discussions, it came up that Arkadi had two daughters, the elder of whom was apparently ill with a little-understood disorder involving her spleen. Her doctors believed she had developed a very rare illness more prevalent to individuals of Mediterranean origin, characterized by a propensity for her white blood cells to die off at an unusually high rate. As a result, these dead cells accumulated in the spleen, causing it to become enlarged and consequently harden. The local doctors recommended that she travel to Italy for treatment as at least there a few of their doctors had achieved some success in battling this disease.

Arkadi, having heard of my healing abilities, asked if I would be willing to examine his daughter and determine if I could help her. I wholeheartedly agreed as anyone can relate to how much we instinctively want to reach out to a suffering child.

The next day I paid a visit to the family at their home. His four-year-old daughter, *Asmik*, was doing what a typical four-year-old might do, if it were bedtime—curled up on the sofa, engrossed in a book with large pictures and print. Her beautiful brown eyes looked tired, and her movements were apparently weak and required deliberate effort. Her mother described her as spending most of her time exhausted on the couch lacking the typical childhood energy to run and play with her friends. I approached the sofa and sat next to the young girl, expressing an interest in the story she was reading. While sitting there, looking on, I assessed her energy level and the condition of her organ systems. The spleen was glaringly the main culprit, which had been the suspected target of her illness. I focused on cleaning the spleen as she sat there contentedly reading her book. I sensed that we had made some progress and, after a few minutes, said I'd be on my way and would check back about *Asmik*.

Later that evening, I had received a call from my friend Leonid, who was very excited about the news he had from his Arkadi. He said just a short while after the conclusion of my healing session with *Asmik* that she appeared immediately transformed. She got up off the sofa, became more talkative, and had suddenly regained her long-lost appetite. Then after stuffing herself, she asked her mother if she could go out and play with her friends! That was music to their ears. Leonid asked that I please return the following day. I was happy to oblige!

The following day and after a third consecutive day, I returned to see little *Asmik* at about the same time. On the third visit, she was already outside playing somewhere in the yard, and her mother had to locate her to bring her in for another treatment.

She responded rapidly to bioenergy treatment, as many children do. After a few more "booster" treatments, there were no further signs of the "Mediterranean illness." To any onlooker, she was just another normal, healthy, active four-year-old, full of energy and delight. Her grateful father, who had to review all exit visas, immediately processed my family's permission to leave the USSR.

Healings on Arbat Street

In order for me to gain permission from the American embassy in Russia to move to America with my family, it took a considerable amount of time to navigate through the complexities and details of the process. With the Embassy physically located in Moscow, we thought it best to move there temporarily to expedite working through the steps in preparation to emigrate. Once we settled into our temporary quarters, for the next two months we waited in line with our papers, going through the daily drill from 8 a.m. until 2 p.m., waiting for our turn, as there were many other families also trying to make arrangements to leave the country. Fortunately, some of our friends were there with us, which made it more pleasant. Not far from the Embassy, where we spent the better half of the day, we discovered Arbat Street, an old shopping district that was being completely renovated and modernized that very year. There were artists, musicians, clowns, and merchants on virtually every street corner. It was known as the "Montmartre of Russia." After a long day at the embassy, we would gather with our family and friends and walk around Arbat Street. Many of the talented street performers appeared to actually be able to earn a decent living by entertaining the passersby. As we observed this, I realized that I didn't have those typical street side talents; however, I could perform a health diagnosis on the spot! That would be a brand new addition to the mix! So we began looking for material to create a makeshift sign. We found a cardboard box and cut out a panel and wrote in large print:

*IN A FEW MINUTES I CAN GIVE YOU
A COMPLETE PICTURE OF YOUR HEALTH!*

We decided that a fair price for a brief diagnostic session would be two rubles. After placing this handmade announcement on a nearby vacant wall, I stood next to it. In only a few minutes, a crowd had begun to gather with interest and inquiry. I began taking one client at a time. The crowd watched with both skepticism and enthusiasm as I evaluated each passerby. The reactions of the subjects were varied, from relief to confirmation, concern, and continued skepticism. Most of the subjects realized, either intellectually or intuitively, that there was an inexplicable level of accuracy in my report. The subjects' body language and facial expressions spoke to the onlookers, who then were drawn in further, wanting to be a part of it all. As the line began to grow, I was able to see about 40 people and then asked those remaining to return the following day. Upon my arrival the following afternoon, a line had already formed. Seeing this, my friends began to have grandiose visions and insisted that I raise the price to three rubles.

One of the subjects in line that day had an interesting situation. He was a man of about 50 who appeared visibly ill. He asked if I might spend a little extra time with him. My heart went out to him. He was from Bulgaria and was apparently one of their leading physicists. He told me that in recent months, he had begun to experience severe headaches. This week, he was sent here to Moscow by his government for testing at the Kremlin Hospital, at a cost of $10,000! He was interested in my diagnostic opinion prior to the formal test, which he would undergo that week. After looking at him for a few minutes, I identified a specific location on his head where my estimation indicated that the circulation seemed inadequate. I believed that this sluggish area could be triggering his headaches. He handed me the three rubles and, before leaving, asked for my telephone number. Sure enough, three days later, the phone rang. It was the physicist from Bulgaria with interesting news. He reported that the Kremlin Hospital's tests revealed that a small section of his head had very poor blood circulation.

We laughed as he said, "We had to pay $10,000 for a diagnosis that you sold me for three rubles."

It was shocking. My girlfriend's family had finally agreed to keep her on life support. Anush was my only hope. As he raised his hands above her, the weak vital signs on the monitor went crazy and all over the place. The medical staff frantically responded in disbelief. After about two weeks of bioenergy sessions, miraculously, Jennifer started to awaken from her coma!

Harold Katz, Studio City, California

CHAPTER 6

Establishing My Healing Practice in the United States

After getting settled into a home in Los Angeles, I very much wanted to establish a healing practice again. To open a business I had to go to city hall for a license. When the agent asked me the nature of my services, I explained to her that I was a healer and could remedy a variety of diseases although I did not have any of the typical certifications. After listening politely, she opened a large book and flipped through the categories of licenses. When she was unable to find a match for my line of work, she called the department head. She too listened to me explain my work and then asked if perhaps I could provide a demonstration, so they could further understand. I was happy to cooperate. She ushered me into a nearby empty office and asked several of her coworkers to accompany us. Coincidentally one was complaining of a headache and another of an upset stomach. I began sending bioenergy to both of them. Everyone watched with curiosity and understandable skepticism. Then to their surprise, within 10 minutes, they were both free of any discomfort. Upon witnessing this, the department head was amazed that without even touching the patients, needing them to disrobe for an exam, or giving them any medications as with other healing modalities, they displayed marked improvements. She stated that she would absolutely grant me a business license under a brand new category, so others could benefit from these healing services. She was so impressed that she said she would gladly provide a reference, should the purpose of the license ever come under question in the future. I was very grateful for her openminded consideration, this was not an everyday request!

After getting the license, I rented a small space in a medical building on Santa Monica Boulevard. I shared an office, including several small examining rooms with a few other doctors, where I was frequently asked to collaborate on their diagnoses. In addition to that, I began to develop my patient base as the word began to rapidly spread of my diagnostic and healing abilities.

Absorbing the Benefits of Someone Else's Bioenergy Treatment!

One of my early Los Angeles cases that was particularly interesting involved a young woman named Natasha, and her boyfriend, Vladimir, who came to my office for a routine checkup. Being immigrants from Russia, they were already familiar with the value and benefit of bioenergy for both the prevention and treatment of disease. I started with Natasha. My evaluation of her revealed a small tumor on her uterus. So I suggested that we schedule subsequent healing sessions to which she anxiously agreed. After my session with her, I turned my attention to her boyfriend, Vladimir, who had accompanied her. We discussed his concerns. He explained that he had a rather severe case of psoriasis, that it had been torturing him for the last several years. He was hopeful that I might be able to help as he had no results thus far with his mainstream treatments. Unfortunately, I admitted that I had no prior experience with psoriasis and thus had no idea if I could have any effect on his condition. I had no idea how responsive it may be to bioenergy therapy. Although obviously disappointed, he didn't comment but did ask if he could continue to accompany his girlfriend at her healing sessions. I was more than happy to have him here for moral support, if nothing else, and to observe. So for the following 10 days, he came with her and sat close by, watching me conduct the treatments. By the tenth day, Natasha's tumor had completely disappeared, and the young couple was ecstatic!

We said our good-byes, and I didn't expect to hear anything more from them.

Two years passed, when out of the blue, I received a phone call from Vladimir. He asked if he could schedule a series of follow-up healing sessions with me because his psoriasis had returned.

A bit confused, I realized that two years had passed; however, to the best of my recollection, I didn't recall attempting to treat his psoriasis. I remembered counseling him and admitting that I was unsure if he would receive any benefit, so I was quite sure we did not treat his condition. He laughingly told me that, evidently, I was unaware of the true range of my abilities. He said that his being present, watching me treat his girlfriend was apparently enough to send his psoriasis into a state of remission for a period of about two years. How wonderful! It is amazing, the power and far-reaching effects of bioenergy, and how this truly demonstrates our connectedness with this universal energy source that surrounds us. So of course, I was delighted to arrange to "treat" him again.

Liver Disorders, Including Cancer, Are Very Responsive to Bioenergy Therapy!

The liver, as critical as it is in our survival, is fortunately very responsive to bioenergy therapy. It's important detoxification function is readily restored, and its associated tumors often disappear quickly, sometimes even after only one treatment session.

One of my early Los Angeles patients was an elderly woman diagnosed with liver cancer—the mother of a very nice young man, Gevork. Her x-rays had revealed a large tumor, and the doctors had immediately scheduled surgery. In this sudden whirlwind of alarming news, Gevork took a step back and contacted me for a second opinion. He knew my reputation from my years as a healer in Armenia and was hesitant about putting his mother through all of the high-tech cancer treatments through which he had seen no survivors. I stopped by his home where his mother was staying. We chatted nostalgically about our earlier days in Armenia, our friends there, etc. During the course of our conversation, I also examined her. I was able to sense her very enlarged and hardened liver. I advised Gevork to postpone the operation so that we could spend some time and reverse her condition. She and her son agreed, and so I began my daily visits and treatments.

After the first few sessions, the associated pain began to subside, and the tumor had begun to soften—both good signs! By the final, 10th session of that first series, the tumor felt as though it were nearly

70 percent reduced in size. That was my sense about her progress, and to check it, I suggested that she return for a follow up diagnostic x-ray. The x-ray results astonished everybody. According to the new x-ray study, the tumor, which was quite large, looked as though it had exploded into numerous small particles. With this evidence of clear progress, she then canceled surgery, took a break from our sessions for the body to stabilize, and then returned to me four weeks later for another course of treatments. This second course of treatments resulted in the complete elimination of the tumor. I continued to monitor her progress for more than 15 years, and her liver had remained clear of any signs of cancer.

As in the psoriasis case mentioned previously, there are often other peripheral health benefits that arise as a result of a bioenergy treatment, even when the target is another organ or individual. Curiously these sessions not only solved her liver issues but also, apparently, eliminated a stomach ulcer and a chronic pain in her left breast. The discomfort in her left breast had sometimes aggravated her to the extent that it even produced complete numbness in her left hand. All of these complaints disappeared as a result of bioenergy therapy targeted to the liver!

There Are No Guarantees in Life—But Bioenergy Gives You Very Favorable Odds!

In my years as a healer, individuals have entered my office in various psychological states and with various attitudes. I've had people come who were already believers in bioenergy healing, and others who were absolute skeptics—feeling they were just grasping at straws. Some of them felt a sense of obligation to submit to this therapy to placate their loved ones. Others have even come with such strong negative preconceptions—perhaps subconsciously motivated—to prove that I could not help them. It was as though their hopelessness had overpowered their normal innate drive to live. I've also encountered family members who have demanded a guarantee of full recovery for patients who had already been declared hopeless or terminal by their traditional medical doctors. I had a case like that in my Los Angeles practice.

An acquaintance of mine, Stepan, introduced me to his boss, Armen, a young man of thirty-five or so, with despondency written across his face. Several months earlier, Armen had begun to experience pain while swallowing food. With each succeeding day, it became more and more difficult to the point where he eventually couldn't consume even liquids. He underwent a series of tests, which revealed a tumor on the esophagus that was blocking the passage of food. The particular location of the tumor on the esophagus made it essentially inoperable. Armen's only option was to consult with a surgeon at the University of Southern California (USC) Medical Center, who specialized in this type of inoperable procedure. Upon contacting this nationally recognized specialist, he found that while the specialist was willing to attempt the surgery, it was extremely dangerous with absolutely no guarantees or certainties, except the cost of the procedure—$250,000!

Armen had a Health Maintenance Organization (HMO) type of medical insurance, which typically requires that the patient use their on-staff physicians and does not pay for such an outside specialist. His family discussed selling their business or their home in order to pay for the surgery. It was at that point that Stepan felt the need to intervene and try to influence them to consider me with my other successful cases as an option for Armen. What did he have to lose!?

Stepan brought Armen to my office, and we spent a significant period of time together. I thoroughly examined him and his esophageal tumor. Fortunately with the bioenergy therapy, we can reach all those areas that the surgical scalpel cannot! Because of the nature of this tumor, I explained that he may need as many as 15 consecutive sessions. Then, depending on his responsiveness, he may possibly require another ten to 15. Each patient was different, so we would monitor his progress and make our decisions as we went forward. Because this was Stepan's boss, instead of my normal fee of $100 at the time, I agreed to reduce it to $50. Surprisingly, Armen insisted that I give him a 100 percent "healing guarantee" before he agreed to the sessions. I found that rather comedic, considering the virtually hopeless options he had been presented thus far.

Stepan was a bit embarrassed, I imagine, and asked me in private to give his boss the full course of treatments and said that he would pay for it himself.

Stepan had very kind intentions, and, not wanting him to overburden him, I decided to treat our skeptical patient at no charge. So Armen agreed to his free sessions. Within just two to three days, the pain that Armen experienced while swallowing had already diminished, and by the fifth or sixth day, he was able to eat and drink without any discomfort whatsoever. By the end of the 10th session, he appeared to be so well that he disappeared and didn't return.

I assumed all was well and ended up forgetting about him for a while.

About a year later, Stepan got back in touch with me and updated me as to what had transpired. After the first series of treatments, Armen, felt well physically; however, his fear and worry that seemed to control his every move were still ever present. That, in combination with pressure from his relatives, convinced him to undergo the operation he originally looked into. Since he couldn't raise the $250,000, he agreed to have his HMO hospital perform the surgery on this inoperable tumor. According to Stepan, when they opened him up, they found nothing. There was absolutely no tumor or blockage to be found in Armen's esophagus!

Bioenergy Bolsters the Immune System—Elimination of Breast Cancer

Robin, a middle-aged woman, arrived at my Los Angeles office with a sizeable tumor in her breast, which had persisted for eight years. She had refused all mainstream medical help in favor of more natural approaches. Over the course of her illness, she had been a patient of an acupuncturist, who was also treating her with herbs. She said that he had developed a unique system to actually measure the overall condition of a person's immune system using a diagnostic chemical testing assay. Although she intended to continue in his care, she was also seeking my help. I agreed to partner in her treatment program. I gave her a brief overview of my bioenergetic treatment approach. She recognized some of the similarities to the energetic concepts that she subscribed to in Chinese medicine, so she was enthusiastic and hopeful that there would be synergistic benefits.

After our first few bioenergy sessions, she reported that the tumor was already becoming softer and smaller. Within just a few days, I also received a call from her very interested acupuncturist. He had examined Robin and was amazed at the rapid change in her tumor. He asked if he could consult with me to learn more about this remarkable progress. I welcomed him to join one of Robin's sessions. During the session, he asked my permission to measure what he termed her "primary immune system characteristics," both before and after the session. He conducted his measurements using a strange technique with which I was completely unfamiliar. First, he placed two boxes on a table and opened them. Inside were several rows of small bottles filled with various colored liquids. Then holding the tips of his fingers in what looked similar to various yoga mudra hand positions, he proceeded to bring the boxes close to certain specific points on the patient's body, each time recording numbers from what he apparently was observing from the bottles. I waited until he was finished gathering this baseline before starting my healing work. Once I started, he observed intently. Upon its conclusion, he again repeated his mysterious procedure with the boxes and the bottles. Finally, he carefully compared the sets of numbers from before and after Robin's session. He looked up at me with a dazed expression. He told me that he would have never fathomed that the activity of the immune system could skyrocket to such an extent and promised to send a printed report of his measurement results. I have no idea how accurate his measurements were and didn't even understand how he arrived at them. But just as he had promised, in a few days I received his report showing exponential improvement after the session: from about a mere 90 "mystery units" prior to the induction of bioenergy to 10,000 units following!

He later informed me that he monitored the residual effects to her immune system by retesting her later that evening. He asked Robin to return to his office so that he could determine how long the effect of the bioenergy would remain in her immune system. Eleven hours after the session, the tests were repeated, demonstrating that her immune system still measured an off-the-chart level of 6,000 units!

Prostate Cancer Patient Enlists My Help for a Remote "Emergency" Healing

Another remarkable case: Several years ago, a friend of mine, Igor, asked me if I'd schedule an appointment for an acquaintance of his, Vahktang, who was visiting from Seattle. Vahktang was a very personable middle-aged gentleman, an accomplished musician, and conductor from Seattle. He was quite well-known both in America and Russia. Recently, his doctors had diagnosed him with prostate cancer and recommended the typical surgery and chemotherapy. Married just a short time with a young daughter, and his career just taking off to include a promising new contract with international performances, this news sent him into a tailspin. Cancer, he feared, could change everything. Igor had told Vahktang about my success with patients, many with very grim prognoses, and insisted that he at least make a point to see me during his stay in Los Angeles. Respecting and greatly valuing Igor's judgment, Vahktang agreed and contacted me right away.

Vahktang arrived at my office for his initial examination and immediately launched into a rapid-fire discussion. "You know I have cancer," he said.

I mumbled something back to him, as I was completely engrossed in my examination and diagnosis.

"Evidently you didn't understand," he continued indignantly. "They discovered cancer."

"That's all right," I assured him. "At least it's just cancer rather than something more dangerous . . . like AIDS."

He looked at me strangely, not certain if my reaction was just casual or uncaring. I assured him that cancer was very responsive to bioenergy treatment and that in my practice, I had many success stories of patients fully recovering from prostate cancer.

I found that his prostate was a bit larger than normal, but by no means was there a need for immediate surgery. I instead suggested we start out with 15 consecutive days of bioenergy treatments. He agreed but asked if it were possible to conduct two sessions a day and achieve the same effectiveness in order to complete the treatment within a week. That approach was frequently effective, so we set out on that course.

By the end of that week, his prostate gland had already returned to normal size. Nevertheless, as he had started out with such a gloom-and-doom belief about his condition, I felt that he might need to ease his mind. So I advised him to consult with a local physician for an evaluation. He agreed to that recommendation and did just that. The following day, after consulting with a recommended local urologist, he stopped by my office beaming. Like a child with a straight-A report card in hand, he proudly showed me an ultrasound study result reporting no abnormalities of the prostate. A manual examination had also corroborated the report and revealed nothing. Vahktang returned home to Seattle just over a week later, filled with renewed energy and a new lease on life, thanks to the encouraging words and advice of his friend Igor!

Over time Vahktang and I became good friends and stayed in touch, continuing our relationship by telephone. With his new contract and the associated international assignments, he ended up spending a year in Korea conducting. Shortly after returning home, he was offered a permanent opportunity to conduct another symphony orchestra on the East Coast. After investigating the area, he loved both the surroundings and that orchestra so much that he bought a house and put down some roots for his family. Periodically, he would call me with details about his house, the city, the weather, and the snow, all of which my family and I missed. We were envious.

On one occasion, instead of our usual chitchat, Vahktang's call was of a more urgent nature. He needed my help for one of his key performers. Only hours away from the opening of a new opera, the lead male soloist found himself in sudden acute back pain, unable to perform—with no substitute soloist available. Apparently, while taking a shower, the soloist, out of nowhere, sustained sudden back pain so severe that he could not straighten his body and was doubled over in pain. Vahktang hoped that perhaps I could treat him with a bioenergy session remotely. It is particularly challenging, yet still entirely possible to connect with someone, even a stranger, from a distance. In this urgent moment, I focused on the soloist continuously for 15 minutes and attempted to connect with him. Fortunately, that's all it took. In that short time, he was able to rise out of his painful, hunched contortion into a straightened position, ready to perform.

Vahktang called me immediately after the premiere and announced that the soloist made it through the entire performance. Remarkably he had rebounded in minutes from his debilitating condition to delivering a fabulous opening night performance—just with a simple 15-minute bioenergy treatment—and from "just a few thousand miles away!

Even Celebrities Were Hearing of My Work!

As the word of my healing abilities was getting around Los Angeles, I received calls from individuals from all walks of life—including celebrities.

It began through an introduction from my friend Yuros, a well-known and accomplished artist to his neighbor, Randy. Randy was a very kind man and was suffering with a rather severe case of multiple sclerosis that left him unable to drive or even walk. Yuros asked passionately if I could make time to work on Randy's condition. I agreed to see him, and in a few short hours, they were already at my doorstep! He began to come to my home on a regular basis, and after only three sessions, he was able to resume driving! Some days I would perform the bioenergy sessions in Randy's home. On one such occasion, Randy introduced me to his neighbors, Bob and Elizabeth Kane. Bob was a famous artist and creator of the cartoon, movie, and television character Batman. Elizabeth was also in the movie business as an actress. Having seen Randy's remarkable improvements in such a short time, they became interested in my healing gifts. Bob had concerns over an elevated prostate screening test level, which can be an indicator of a risk for prostate cancer. Elizabeth also had a breast lump that she wanted to eliminate without invasive procedures. So Bob and Elizabeth invited me to visit them daily to see if I could shift their conditions to a more healthy state.

Their apartment was wonderful, full of Bob's Batman drawings and other paraphernalia. Over time Bob's prostate tests dropped back into the normal range, and Elizabeth's lump disappeared. We became good friends, and they called me for "tune-ups" if they felt like anything was out of balance—before it became anything serious.

One afternoon, to my surprise, Bob took one of his drawings and signed it and presented it to me!

We continued to stay in touch, and in fact, they attempted to contact me while I myself was recovering from a car accident. Elizabeth, not knowing that I was recuperating, called my home looking for help, as Bob was in the hospital. Fortunately, Gia, my bioenergy teacher, was in Los Angeles at the time, so he intervened in my absence. Bob bounced back quickly and was soon discharged. Once I was back on my feet, Bob and Elizabeth were kind enough to invite Elada and me to lunch on Rodeo Drive. It was a very exciting experience for us!

Soon after meeting Bob and Elizabeth, Brenda Richie, Lionel Richie's then-wife, called me. Brenda was looking for other options for her father, who was diagnosed with pancreatic cancer, and asked that I treat him using bioenergy. In combination with bioenergy and other natural approaches, his improved. On one occasion at the Richie's home, Lionel expressed some curiosity about the bioenergetic techniques so I did an assessment of him. His energy was very balanced and flowing. I found him to be in excellent health and he didn't need me!

My daughter, Arevik, and I ran into Lionel at the airport a few years back. After some reminiscing, he was kind enough to send us this souvenir photo!

(Note: I have left out the details of this case as Brenda Richie did not have an opportunity before publication of this book to review my recounting of her father's treatment. My hope is that her testimonial can be included in a future publication).

At 19, Jennifer in "an irreversible coma"

One of the first articles about my Los Angeles healing cases had been published in the local Armenian newspaper *Paros* and subsequently in the Russian publication *Panorama*. The word began to spread in the Russian community, as many of them knew about bioenergy therapy from Russia. As a result, the number of my Russian patients increased.

I received a frantic phone call early one morning from a Russian woman, Galya Sarnoff, who had read an article about me in a Russian newspaper. Her son Harold's girlfriend Jennifer, 19, had been in a coma for about two weeks. On the day that she lost consciousness, Harold recounted that Jennifer, had not reported to school that day. The night before he had taken Jennifer and her friend to dinner and then had dropped her off at her mother's home. So this was very unusual. He called her house to check on her, and when there was no answer, he became concerned and rushed over to her home. Jennifer had been rushed to the local hospital emergency room in a coma. According to her mother, she was fine the evening prior. She and her mother were laughing and joking that evening. Everything about Jennifer appeared perfectly fine. She was a typical happy teenage girl in love. The next morning, seemed to have overslept, but mother was unable to wake her. Her pulse and respirations were faint. Her mother immediately called 911. Paramedics arrived quickly and were able to restore a very weak heartbeat using electroshock apparatus. They then transported her as quickly as possible to the nearby Burbank Hospital, where she was immediately connected to cardiac and respiratory support.

Jennifer had an unknown cardiac weakness from childhood, and suffered a heart attack, followed by a stroke. She had then slipped into a coma. She remained unconscious, connected to monitors revealing an absence of normal brain-wave patterns for the next several weeks—her family and the doctors looking on, feeling powerless. In the doctors' estimations, Jennifer's condition was hopeless. They said that based on the monitor readings, she would not be able to recover, and they recommended that the family agree to remove her life support. Harold had not left her side and refused to accept the fact that they wanted to give up on her. He became extremely angry, insisting that Jennifer's parents reconsider. The boy's love for their daughter was moving. In their compassion, Jennifer's parents agreed to step back and delay any further actions. Harold had no solutions at this point—just a little more time. He prayed a heartfelt prayer asking for God's hand in bringing Jennifer back to them.

Harold's mother, Galya, was aware of her son's seemingly imminent loss and had also heard of my bioenergy work. So with the chaos that had taken over Harold's world, she remembered the article about my bioenergy healing and made an "emergency call" to me. By the time she reached me, it was 10 p.m. She believed the doctors were planning to remove life support the following day. Under the emergent circumstances, I volunteered to go to the hospital that evening. Harold met me there. He asked the nurses to give us privacy and told them I was a friend of the family, who had come to pray for her. Jennifer was clearly in a very deep coma, barely breathing without the help of the respirator. There was still no measurable brain function. According to the attending physicians, because her heart had stopped that morning at home while she slept, her brain had been deprived of oxygen for a long enough period of time, such that even if she became conscious, her brain would be irreversibly damaged.

I rapidly began to scan her to determine my diagnosis and an immediate course or action. While I recognized her condition was very serious and challenging, my instincts told me that the damage could be repaired. The attending staff had left us quietly in the room. I began by focusing bioenergy to her head and brain and then continued by restoring the flow of energy to all of her body's energetic channels. I intuitively felt her responding, and the electronic monitors responded chaotically to the energy in the room as well. The nurses ran into the room trying to determine why they were "malfunctioning" with the alarms going off.

After spending some time with Jennifer, we took a break so that I could be introduced to her mother. She had heard of my work and, although skeptical, was grateful for the hope I brought. She agreed to convince the doctors to further postpone the disconnecting of life support so that we could bring her back.

I returned to the hospital the next morning. For the first time, there was noticeable movement in Jennifer's hands and legs. Knowing that comatose patients still have awareness and can still sense the world around them, I decided to incorporate hypnotic commands into our treatment.

Jennifer responded perfectly to orders, such as "Move just your left hand,. Now your right hand. Move just your left foot," and so on. Jennifer was by no means brain-dead!

Harold remained in the room. As Jennifer spoke English, and my English was still a bit broken, I "commanded" her to listen and follow the voice and the instructions given by her boyfriend. It was a perfect collaboration. While I was busy focusing on my bioenergy treatment, Harold could issue the hypnotic suggestions and "entertain" her by reminiscing about happy events and memories in their lives together. Jennifer, although still technically in a coma, responded with attention and even smiled at the humorous stories. Each day she demonstrated more movement, strength, and awareness, until about two weeks after my initial visit, she finally awoke.

Her doctors were dumbfounded, unable to offer any explanation other than calling it a miracle.

Jennifer's Coma Recovery Gives Hope to Another Child

In 1997, I wrote an article in a local Persian newspaper about bioenergy healing. The article generated a lot of interest from the Persian community that resulted in many new patients for me. I was just beginning to get my strength back after a life threatening automobile accident of which I will elaborate elsewhere in this book. I received a call from a former patient of mine, Azura, a Persian woman who had seen the article. that I had used bioenergy to bring an individual out of a comatose state. The night before, she had seen a Persian television program where a man from San Diego was on the air desperately asking for any help for his eight-year-old son's condition. The young boy had been hit by a car while riding his bicycle and hit his head on the car windshield. He was rushed to a nearby hospital; however, the force on impact had been so great that he fell into a deep coma. After monitoring him for several days, the doctors determined that his brain was no longer alive and asked for the parents' permission to discontinue all life support. The parents were distraught and said they needed time to think about the doctors' recommendations. That same day, the boy's father decided to come on the air and beg the Persian community for any ideas, modalities, or facilities that could save their son's life. The father's poignant plea touched the hearts of the audience. Upon seeing this, *Azura* recalled the story of Jennifer who had also been in a coma and had been saved by her young boyfriend's insistence that they not give up on her and my intervening support. *Azura* asked me if I could go to San Diego and stay for 10 days to see if I could help revive him. I explained to her that I was just recovering from a coma myself and would not be able to travel that far from home for an extended period of time. She continued to ask if there were something I could do as this young boy's plight, and the father's will to save him had taken ahold of the community. I finally agreed to attempt to travel there for one day and return.

The next morning, Elada drove me from Los Angeles to San Diego. I saw the young man and knew he would need more than one session to bring him around, which I would not be able to perform under the circumstances of my recuperating condition. As you may recall, in Jennifer's case, although she was in a coma state, her subconscious was aware and alert and was receiving and responding to the words and other stimuli around her. It's essentially like being in a hypnotic, very suggestible state. So using verbal commands, we can communicate with the comatose patient and bring them back to consciousness. With this in mind, I used hypnosis with the young boy, *Cyrus*, who was in this very open and suggestible comatose state. I left him with the suggestion that his father would be speaking to him and to pay attention and obey his father's every word. I gave *Cyrus* a suggestion to follow any commands that his father would give him. I explained to his father that he could use his calm and supportive voice to coax his son into knowing that he could heal and that he would awaken. Every day and every chance he could, the father lovingly spoke to his son, telling him that they loved him and that they were ready for him when he felt well enough to come back. Little by little, there were signs of movement and brain activity, until one miraculous day, less than one month later, little *Cyrus*'s eyes opened to the welcoming arms and joyful tears of his parents.

We Are More than Just This Physical Body

Jennifer and *Cyrus*'s story further emphasizes that we are more than just a physical entity with bodily functions. The electronic apparatus that are attached to a coma patient's body is limited to measuring physical phenomena, and those readings are all that our medical establishment can use to base their understanding of health and their prognosis for patient outcomes. However, in actuality, we have four "bodies," with the physical body being the first and most primitive. The comatose state is essentially a shutting down of the components that run that physical body. Thus we still have full use of the other three bodies known as our ethereal, astral, and mental bodies. In the absence of a fully functioning physical body, these other three bodies can continue to function normally. (I will provide more information about these other three bodies in Chapter 7.) Comatose patients have awareness and comprehension of everything going on around them. However, they may not be capable of responding in the physical manner that we are accustomed to. Yet it is evident, especially after receiving a hypnotic order or request that he or she will attempt to respond in some way. Since the comatose patient's physical consciousness is "turned off," logic dictates that the practitioner's work is on the subconscious level, such that the other three still-functioning bodies can be reached and enlisted.

Using hypnosis, it is possible to guide a patient's current state of consciousness into a hypnotic trance. This is exactly what I was able to achieve with Jennifer. As mentioned, part of my earlier healing training consisted of the mastery of hypnosis. I had achieved a high level of mastery in hypnosis even prior to entering into bioenergy work. I am able to very rapidly command most subjects to fall "asleep" virtually instantaneously. When Harold and I worked together, he told Jennifer stories that she would remember and that would stimulate her mind. One day, one of his stories actually brought tears to Jennifer's eyes. Yet the hospital's electronic monitoring devices continued to register a complete absence of brain activity—"brain dead," as it had no way of measuring those processes occurring beyond the physical level to which she was clearly responding.

How then, I ask you, can we make life-and-death decisions based upon these inadequate instruments?

Three Cases—One Family

Gretchen, a former patient, contacted me a year later in 1997 to set up an appointment. She had come to me in the past for testing anxiety, which is another one of the diverse issues that can also be addressed with bioenergy healing. She was bright and had graduated from law school with flying colors; however, she had been unable to pass the bar exam for two consecutive years. She had an excellent grasp of the subject matter even as complicated as it was. Her challenge was that there was a time constraint, and it was essential that she answer the questions in an efficient manner. This pressure preoccupied her, affecting her ability to focus and manage her time, and resulted in two failed attempts. So our work together was aimed at calming her anxiety using bioenergy and hypnosis. I also used suggestions so that she was better able to prepare for the exam. As *Gretchen* studied for the exam, she was able to absorb every word of the text verbatim as though she had a photographic memory. On her next attempt to pass the bar exam, she not only passed, she completed the test in record time! This was absolutely remarkable. This time she was able to tune out all distractions and remember only the words in her textbooks. It was as though she was in another world. She was oblivious to all of the voices and sounds around her. Nothing would get in the way of her concentration. She was such a perfect candidate for hypnotic suggestion that she followed the voluntary commands precisely and passed the test effortlessly. She returned to the LA area and opened her law office in Glendale.

Knowing the power of bioenergy to do what seemed to her to be the impossible, *Gretchen* contacted me again, this time on behalf of her aunt *Margaret*. The doctors had diagnosed *Margaret* with colon cancer and wanted to perform surgery to remove 20 centimeters of her intestines, which they considered to be the area of the tumor. *Gretchen's* aunt opted to try bioenergy therapy first and came to my office for 10 sessions. She had felt considerably better after that course so I advised her to take a break and repeat the same treatment plan in a month. Unfortunately, at the insistence of her worried and skeptical relatives, even with her marked evidence of improvement, she was encouraged to undergo the surgery.

I received an update on her progress and heard that when they opened her abdomen they were astonished because only about four centimeters or 20 percent of the original 20 centimeters of cancerous tissue remained! So as a result, instead of cutting 20 centimeters of her intestine, they took only four centimeters! Who knows if we had the other subsequent sessions if there would have been anything malignant remaining?

This story goes on. *Margaret's* husband, *Robert*, who had joined them at the sessions as an onlooker, was intrigued with how both of them with such distinctive conditions had fared so well.

He was a very healthy and energetic man of about 57. While *Robert* was at my office, he decided to show me a large mole on his upper chest, just below his left shoulder. He reported that the mole would occasionally bleed if scratched or touched roughly. He asked me if I thought he should have it removed. I advised him, in my opinion (as I was not a medical doctor), to give it some time and simply be attentive to it for a while. I suggested he try not to touch it unnecessarily and, in case it were to bleed, to simply use his saliva since that is the best cleaning and antibacterial substance. I said that even if this irritation should worsen and the mole should grow, there are simple nonsurgical means to remove it without any risks. I further explained to him that surgical removal could actually stimulate the development of cancer, which could spread to the lymphatic system.

Several days later, he returned with a bandage, showing me that he had opted for the mole to be surgically removed. He showed me the location, which was a surprisingly large and deep wound. He explained that the follow-up medical tests had indicated the mole tissue was not cancerous and asked if I could simply use bioenergy to clean the wound. At that time, I advised him to come back to me for several additional bioenergy sessions, just in case the medical diagnosis proved wrong. Biopsies tend to be inconclusive because there could be a mixture of malignant and benign cells at the site and, depending on which cells are examined, a different diagnosis could result. Furthermore, I felt that additional bioenergy treatments could help prevent any future possibility of cancer developing in the area. Nevertheless, he chose not to return at that time.

Two years or so later, I received a phone call from *Margaret*. She was distraught and informed me *Robert* was in very bad condition, and they needed my help. When I asked why they didn't call me sooner, she said that they were uncomfortable about calling me because they had ignored my advice from two years ago. I know that the medical establishment is well meaning, and many patients also believe that their doctors are infallible. So I understood their past decisions. I welcomed them to come back for further treatments to see what I could do. That same evening, they arrived at my house; they brought me up to date on what had transpired after the "simple mole removal" of two years ago. Apparently, when *Robert's* mole was removed, the lymph nodes under his left arm became inflamed. The doctors then recommended removal of the lymph nodes, so *Robert* complied. For some time, everything appeared to be normal. But then he started to experience severe headaches. Tests and x-rays were now revealing a large tumor and several smaller ones in his skull. They performed an initial surgical procedure to remove the largest of the tumors. The others were not easily accessible, so they were subsequently removed using radiation. Several months after these procedures were performed, *Margaret* brought him back to me, looking for a miracle for an already exhausted body. *Robert* was still experiencing severe headaches and was having difficulty walking and talking. and he had a large Y-shaped surgical scar visible on the right side of his head, which was also still quite swollen. He was very frightened and depressed.

After several sessions with me, Robert' speech improved dramatically. After 10 sessions, the swelling on the right side of his head subsided. He began to walk more normally and appeared less fearful. We were making tremendous progress., I myself was then in a life-threatening car accident and was unable to continue *Robert's* treatments. During my recuperation, *Robert* opted to go back to his conventional treatment. He deteriorated and died.

Seeing Isn't Always Believing

So far in this book, I've described a number of successful cases involving patients who had a keen awareness and were in touch with their bodies, able to sense the healing process—as inexplicable as it may have been to their physicians or others around them. At the same time, I've had cases where, although the bioenergy treatment is proving to be effective, the patient's surrounding support system and other circumstances changes the ultimate outcome, sending it off course. This type of divergence occurred in this early case in Los Angeles:

I was asked to come to the hospital to consult with *Lena,* 20, diagnosed with liver cancer. *Lena's* parents were both dentists, and several of her other relatives were also doctors. I make this point because the preexisting medical orientation of the family members in favor of mainstream medicine can often cloud the real improvements that a patient experiences during bioenergy treatment. *Lena* had been experiencing pain in the liver region along with the typical loss of appetite and weight loss. Her concerned parents immediately admitted her to Cedars-Sinai Hospital for testing. The battery of studies revealed a malignant liver tumor.

I was accompanied to the hospital by one of her relatives, who also supplied me with *Lena's* complete medical history. The documentation indicated that she had been scheduled for an operation; however, it had been postponed. Because of *Lena's* progressing weakness, discouraging blood tests, and an apparent loss of will to live, the procedure was delayed in the hopes of waiting for her to be in a more suitable physical and mental state to undergo an operation. Upon entering the ward, I saw *Lena* in bed with her parents attending to her. The family had contacted me, not expecting a cure at this point, but to simply try to improve her energy state and appetite, and, to perhaps help lift her depression.

I greeted *Lena*. Her parents stepped aside so that I could examine her. After spending a few minutes with her, I exited the room to speak to them about my assessment. Based on my perceptions of her condition, I was confident that I could produce some marked improvement bioenergetically. By focusing on the cleaning and restoration of the energy flow in the liver, I felt certain that we could improve her strength and possibly even improve her psychological state as well. *Lena's* parents, were skeptical but agreed to have me return for regular sessions.

I began *Lena's* first bioenergy session with several visitors in the room, curiously observing. I'm sure they didn't know what to think as they watched me waving my hands a few inches away from her body. After I finished and satisfied with that first treatment, I asked the now very mystified visitors if they would give me a few minutes privately with my patient. When her visitors had left, *Lena* began sobbing, expressing her fears, and feelings of grief and hopelessness. I responded encouragingly about her body's ability to rebound, and told her of my many patients who had initially similar, seemingly hopeless outcomes. I said, "There is so much healing power in your body that you have at your disposal that even the best doctors can't begin to understand." Her eyes began to show signs of life, and she felt more optimistic about her possibilities for recovery.

Within four days of treatments, *Lena* was feeling and looking noticeably better. By the seventh day, her amazed parents had her released from the hospital so that she could continue to recuperate in the comfort of her home. They insisted that I continue the treatments at their home. *Lena's* condition continued to improve. She was eating better, the yellowish tinge of her skin was fading, and she looked rejuvenated. She was now spending the better part of the day out of bed walking around.

After two weeks of treatments, she began going for walks again with her fiancé. By that time, I recommended that it was time for us to take a treatment break for about three weeks for her body to do some of its own healing now that her energy channels were restored, and her immune system could rebound.

After the three-week hiatus, I suggested that we resume with another course of treatments, until such time that we detected the absence of the cancer.

At this point *Lena* expressed that she was feeling wonderful and never imagined that she would have ever felt so much like herself again. Her parents and her relatives, witnessing the return of her energy, vitality, and normal lifestyle, were still struggling with the acceptance of this nontraditional recovery. Many of them being doctors, the change in *Lena*, sans high-tech medical intervention, contradicted their understanding of the disease process.

Finally, on what I enthusiastically determined to be the last day of required treatment, I announced to her father the wonderful news that *Lena's* cancer had completely disappeared. No operation would be necessary as there was no longer any trace!

Lena, although she had actually grown to welcome our visits in many ways, was overjoyed to have her life back. She and her fiancé got married and were ready to celebrate her newfound health and their life ahead. And they didn't waste much time, as they booked a trip and jumped on a plane for Hawaii.

I thought, "What a happy ending. This is what makes my work so fulfilling beyond words . . ."

Three Months Later

Three months later, one of my other patients, who knew of *Lena*, casually mentioned that she had heard about *Lena* undergoing cancer surgery. I was shocked. How was this possible? She was clear of cancer. I was completely rattled at this rumor. After making some immediate inquiries, I learned that she had returned from her Hawaiian vacation, rested, strong, and healthy. So what could have possibly been the reasoning? Discouragingly, with more investigation, I learned that her father had begun feeling the pressure from their other family members in the medical profession who cautioned him to "not sit back and play with *Lena's* life." How, they argued, could her father—in the health care field himself—trust the life of his only daughter to some mysterious healer, when all of the medical studies clearly showed that she had liver cancer? "Sure it is wonderful," they said, "that she feels better—all the more reason now for her to have the surgery that she once didn't have the strength to endure." *Lena's* father, succumbing to the pressure, convinced *Lena* to have the surgery against her will and instincts.

Lena was operated on by one of the shining stars of Cedars-Sinai Hospital, who also happened to be the doctor who had performed the original diagnostic biopsy... The doctor, an alternative medical physician who knew *Lena's* father, spoke to him after *Lena's* death. *Lena's* father said that when this top-notch surgeon opened *Lena* up and saw the condition of her liver, he was struck by the fact that it was virtually normal, admitting later that there was no real need for him to have intervened. performed surgery. However, during surgery, the accidentally severed *Lena's* liver, and inadvertently also damaged her gallbladder. When *Lena* came out of surgery, there was a great deal of inflammation in the area, as well as a myriad of other complications. Later, the doctor had to perform two subsequent corrective surgeries, which were essentially unsuccessful. *Lena* never recovered and died a few months later.

One Neck Tumor Patient Refers Another!

Nine years after her son Laert's healing from his neck tumor, Anahid called me on behalf of her friend Julia, who had a similar walnut-sized tumor—this one was on the right side of his neck. Once again, the doctors recommended radiation and chemotherapy, and then surgery. Fortunately, knowing Laert's doctors had made the same radical recommendations and knowing Laert's ultimate success story by taking a more holistic path, she refused that course of action. Instead, she returned from the hospital and contacted Anahid immediately for advice on how to reach me. Anahid reached me immediately, and we got her over to my home the following day. After just a few days of treatment, the tumor began to soften. At the end of the first 10 days of bioenergy treatment, the tumor shrunk to half of its original size. Then after a second course of 10 treatments, the tumor totally disappeared. Interestingly enough, Julia was open to using bioenergy treatment and ultimately got her MD and opened her practice in Montebello. How wonderful it would be if more traditionally trained doctors would see the value in bioenergy and could recommend and integrate it into their practices.

Are Cellular Phones Causing Brain Tumors?

Although a patient can have a full recovery from cancer, as did Laert with his neck tumor, there are still environmental causes of cancer that continue to be revealed that often are unknown as a threat to our health until after the fact. In 2001, Laert was now twenty-five years old, a strapping, handsome young man. Out of the blue, he began to experience severe headaches that had gotten so extreme that his family had no other choice but to seek relief for him by taking him to the hospital. They admitted him for tests and immediately put him on morphine to subdue the pain. I received a distressed call from his mother, Anahid, who said that the medical studies revealed a large tumor in the left side of his brain. The doctors were advising them to submit to the typical three-pronged treatment consisting of radiation, chemotherapy, and surgery. Having heard this before, when he had the neck tumor, she called to discuss the situation with me. Feeling the urgency in her voice, I wrapped up what I was doing and arranged to be at the hospital in just a few hours.

While at the hospital, I asked to see the brain study results. The medical staff showed us a computer-generated image of the left side of the brain, which revealed a several-inch tumor that spread from the forehead to the back of the head. With this detailed imaging, I was able to see the location of the tumor more precisely. I started to perform bioenergy treatments focused on the tumor site. Laert showed some progress that day as he was able to open his eyes. I agreed to return over the course of the next few days. With each subsequent treatment, he became more active, and the headaches had clearly subsided. He began to regain his talkative nature, conversing with me, his relatives, and friends via cellular phone. The use of the cellular phone, I felt, was not in his best interests, as I believe it may have been the very cause of this tumor that was located just where he held the phone. I queried his mother as to how long he had a cell phone and his usage level. She admitted that sometimes he used it for more than ten hours a day and always on his left ear. Even after having long nightly conversations, he slept with it under his pillow.

One day, when I was performing a session on Laert at the hospital, the doctors were reviewing another radiological study that had been performed earlier. We followed him to the radiology department to discuss the findings together. There were young interns with his x-rays in hand. They were absorbed in reviewing and discussing the films. Finally, they had drawn their conclusions and solemnly stated that it was essential that they irradiate the optical center of the brain in order to prevent future blindness. This recommendation raised great concern to me. On behalf of the family, I asked the young doctors how they could be certain of the accuracy of the x-ray beam as it would be difficult to pinpoint their intended location without compromising another important brain center. I also pointed out that the radioactive beam would travel through other parts of the body and could have other unknown and likely destructive effects. They admitted that they did not know the full extent of the side effects of the procedure; however, their professor was on his way and could address all of our questions. The professor did arrive shortly as promised to review the radiological studies. He concurred with the interns' recommended treatment plan of radiating the optic nerve. Unfortunately, he did not have any clear answers for how Laert's body would be protected from unnecessary radiation exposure to other critical centers and functions of the brain and body. We asked what the process was to refuse the treatment. They indicated that Laert's signature was all that was needed to leave the radiology department and forgo their recommended protocol. We continued to perform the bioenergy treatments, and several days later, he was still headache free. Optimistically, his relatives asked for a follow-up radiology study of the brain to see if there was any evidence of improvement in his condition. They agreed and the results were unexplainable. There was no evidence of any tumor left in Laert's brain. The hospital staff was speechless. Finally, they decided that it must have been the pharmaceuticals that were being administered to Laert that dissolved the tumor. Even so, they warned that there was a great chance that the malignancy would return, which would require another surgical procedure, along with the insertion of a localized chemotherapy capsule. This capsule would theoretically deliver a continuous small dosage of poisonous chemicals to destroy the potential for any new cancer cells to grow. Fortunately, in this case, Laert's family refused to acknowledge the doctors' fears and decided against any further surgeries and chemotherapy protocols. They all went home feeling grateful for his recovery.

Two days later, Laert's mother called me and asked that I visit them at their home. I came right over that evening to find that Laert was in great pain; however, this time it was generalized across the entire body. He was unable to sleep or eat or even engage in any conversation with his relatives. The pain persisted around the clock without ever subsiding. I explained to them that because he had received morphine in the hospital for such an extended period of time, his body had become physically addicted. As a result, it was responding with pain as a request for more morphine to quell the addiction. He would have to be patient to get this narcotic out of his system. So Laert's mother took him to the hospital to see what could be done to help with the transition and wean him off this addiction. Instead, the hospital said they would only readmit him if he agreed to the surgical procedure to insert the chemotherapy capsule, as that was his last "prescription" from the doctors. Feeling helpless, and even though there was no tumor that remained, Laert's mother agreed to the costly and unnecessary surgery to implant this capsule of poisonous chemicals. In my experience I believe that these cancer-killing chemicals are actually counterproductive, as they destroy much more than cancer cells in their path and cause further compromise of the body's immune system and self-healing potential. It was a high price to pay with so many potential side effects, when all that was requested was help to undo the addictive effects of morphine.

Shogik's Low Levels of Blood Platelets Respond to Bioenergy Healing

In early 2007, I received a call from my friend Harut, who owned the very reputable A&A Sign Company, a well-known company known for its quality work. Harut and I had known each another for almost twenty years, and he had provided a lot of support to me when my family had arrived in America. At this time, he reached out to me for help. He shared with me his concern for his wife Shogik's health. He asked if I would see her, and so I insisted they come over as soon as possible. Within a few hours, they arrived at our home.

Shogik had been working at a drugstore and kept long hours, leaving home early in the morning and returning very late at night. She had visited her doctor a month before for routine blood work, which came back with some abnormal readings, including a dangerously low platelet blood count. They immediately put her on prednisone and told her she may need surgery on her spleen to correct an autoimmune disorder that was causing her platelets to be attacked. Unfortunately, she was very upset about the side effects of the prednisone, which included a twenty-pound weight gain and such extreme swelling in her face that she could hardly open her eyes. I did a quick assessment on my own and felt she didn't actually have any serious conditions. Her liver, spleen, and kidney were "heavy"; however, that was simply a result of the prescribed prednisone.

It is a powerful drug that essentially destroys all organ systems and bones that were already getting weak and fragile. You may recall the section about Zara, the lupus patient, who too had been prescribed prednisone, which caused osteoporosis.

I asked Shogik a few questions about her diet and her habits and discovered that she was not eating well. Basically, she was on a diet of on-the-go food. It was a rare occasion that she would consume fruits and vegetables even though she fortunately had a big backyard with many fruit trees accessible to her. I encouraged her to eat more unprocessed foods as she could get them even from her backyard. I recommended that she immediately discontinue the use of the prednisone and any other drugs. To support her healing, I taught her a few very important yoga exercises and offered to give her a series of bioenergy treatments. She was thrilled to have the potential to heal without the need for harmful drugs. She came to my home, and I worked with her, performing regular bioenergy treatments. After four to five days of treatment, she already felt like herself again! Her hematologist, not knowing that she had discontinued her prednisone, reported that her platelet levels had returned to normal. Shogik's blood pressure and her sleep also improved as an unexpected bonus!

Garnik Is Healed Using Bioenergy at a Distance

In the winter of 2007, I received a phone call from Moscow. It was a kindhearted female voice, who introduced herself as the neighbor of Bobken, my wife's cousin. She explained that a short time ago, her son, Garnik, had begun to experience terrible headaches. After several medical checkups, her physicians found a large brain tumor. Shocked and in disbelief, they sought a second opinion from the specialists in Moscow. The diagnosis was corroborated, and they immediately admitted the boy to the Moscow hospital. The physicians were preparing to perform surgery, radiotherapy, and chemotherapy as a treatment for the tumor. Garnik's mother refused. Instead she tried to make contact with one very famous Russian healer and television personality known as Allan Chumak. She did manage to reach Chumak; however, sadly, he said he was unable to perform a long-distance healing and instead requested that she bring her son to him. It was not feasible for them to travel. Still seeking help from a healer, they heard about "Anush," who was able to perform distance healing; he was now living in the USA.

The family began to put out feelers to see who may know how to reach me. How remarkable that my wife's cousin just happened to be one of their neighbors! As a result, we were quickly introduced by telephone. Naturally, I was glad to be of service to and collected the necessary information to perform the remote healings.

I was able to sense the significance of the tumor even at that distance. Even though it appeared to be well established, after just a few days of concentrated effort, the tumor began to shrink. As we continued the healings, the pressure on his head was reduced and the boy's headaches began to subside. With the obvious improvement trend, she took Garnik out of the hospital and brought him home. They called me from their home in Yerevan, Armenia, and we continued the remote course of treatment. After a second course of distant healings, the tumor continued to shrink until it had disappeared completely!

In the winter of 2008, a year later, Garnik's mother came to Los Angeles to visit her daughter and brought me a photo of her healthy son. Having courageously substituted the prescribed exhausting mainstream treatment protocol with an immuno-supportive course of bioenergy treatments, his body responded and he got his life back. He was healthy and energetic, just as a young man of his age deserved to be!

Susan Moss, Author of *Keep Your Breasts*, Also Healed Herself Using Bioenergy

I received a call from for an appointment from author, Susan Moss. She had heard that I could heal tumors and was specifically interested in seeing me work on breast tumors. She had heard reports that using bioenergy, I could influence breast tumors to soften and shrink sometimes just within minutes. She wanted to meet me and learn more about the technique and hopefully witness this with her eyes. It was apparently her lucky day as the day she came to meet me, I coincidentally had a breast cancer patient. Susan introduced herself as a writer who wanted to do a book specifically about breast cancer. She wanted her book to be a communication vehicle for treatments that were free of drugs, radiation, and surgery—and in the case of bioenergy, the treatment includes on direct contact with the tumor.

Susan herself had gone through her healing path of refusing mainstream treatment of a tumor that was literally rotting through her skin. To her credit, she had created her regimen of a healthy lifestyle and using the power of her mind and bioenergy to eliminate the tumor within three months—a deadline she had set for herself, which was the date her doctor wanted to do surgery. When I met her, I was able to detect the energetic remnants that indicated that there had been an issue in that region of her body, yet she had obviously resolved it on her own.

After a short time, I received her published book in the mail. She thanked me for the demonstration of my techniques and included a discussion of bioenergy in chapter 7 of her book. We have since kept the lines of communication open, and I have received other contacts from patients—and others interested in bioenergy—who have heard of me through Susan's impactful and eye-opening book. I highly recommend it.

While in the process of waiting three days for medical blood test results, Anush tuned into my son's condition remotely and sensed that in fact he needed emergency surgery for appendicitis. Thanks to Anush's insistence, the doctors said we got to the hospital "right on time" and prevented my son's appendix from bursting!

Karina Yeproyan, Glendale, California

CHAPTER 7

Cancer—Does the Medical Establishment Need a New Approach?

Strides in Cancer Treatment

According to the World Health Organization (WHO), cancer accounted for more than 7.6 million deaths in 2008. It is estimated that there will be eleven million new cases each year by 2030 (World Health Organization Web site, June 2011). The American Cancer Society reports that over 569,000 Americans will die of cancer this year. That's more than 1,500 people per day. What many people would probably be even more surprised to hear is that the death rate has been climbing since the 1930s (American Cancer Society—Cancer Statistics Presentation 2009). Such statistics are not only extremely discouraging but also mind-boggling. Why, with all of the marvelous scientific and technical progress we read about daily, are we not seeing a more dramatic reduction in the numbers of people suffering and dying? Perhaps we need to rethink our approach.

During my twenty years in America, I've encountered many cancer patients and have learned a great deal about the details of the medical history of their illness and their struggle to recover. It seems to me that the techniques used for cancer diagnosis are somewhat effective, yet the treatment benefits have proven to be short lived. I received a letter once from the University of Southern California (USC), the location of the well-respected cancer treatment center known as the City of Hope. In their letter, they asked for financial support and touted information about their achievements in the battle against cancer. They claimed some degree of credit for their role in extending a patient's life expectancy following cancer surgery, "from five to now seven years."

Another Perspective to Consider

From my perspective, I didn't believe this was truly an achievement as, in my opinion, any woman having a small tumor in her breast (as an example) would generally live that long (five to seven years) or longer without any surgical intervention. When a biopsy report indicates the appearance of cancer cells, this is not a cause for panic or any rash decisions. First of all, laboratory reports are too frequently wrong. There are varying statistics on this subject. I've heard that as many as 30 percent of such reports are incomplete or inaccurate. Second, cancer cells can exist in our body at any time, and our immune system routinely disposes of them. And they may temporarily pass through our lymphatic system. Often there are tests conducted for cancer in the lymphatic system, yet these cells that are found may be "on their way out" through an immune system process and filtration through the lymph nodes.

Our lymphatic system is a link between our physical and ethereal bodies, and we believe that cancer originates because there is some problem in that energetic body. That is why the cure has to be energetically—not physically—based. From a physical standpoint, cancer cells may have the potential to linger in the lymphatic system because the ability of the blood to "wash" that area and self-heal is not as efficient in that area as in other parts of the body. This is part of the basis for the therapeutic nature of the yoga positions. The bends created by the yoga asanas or "poses" increase the blood flow to a particular area and the ability for that area to be "washed" by the blood. However, once those cancer cells are outside of the lymphatic system, they are destroyed by our immune system. Even if cancer cells do settle into our lymphatic system, the potential problems are of no comparison to the interference of proper functioning and the body's overall weakening brought on by surgery to that region—ironically exacerbating the condition it is intended to correct. This may sound like a scary or foreign concept as we've all heard and probably believe the term that the patient is "all clear" of their cancer. There is really no such condition that our medical tests can detect. Cancer cells are roaming around all the time. It's up to our immune system to keep the upper hand and to keep ourselves healthy bioenergetically.

Psychological Impacts of a Cancer Diagnosis

It is startling to me that in America, a patient can be formally diagnosed with cancer with little more evidence than a medical suspicion. Upon hearing this news, the patient's psychological response can have as much to do with the outcome as anything else. Some patients are fortunately psychologically strong-willed and recognize that they ultimately retain the control over their body and their health. Other individuals may be unable to handle such devastating news and feel at the mercy of the skills of their authoritative physicians and what the medical system has to offer. The news can come as such a shock to their system that they become distraught, consequently weakening their body's ability to fight back.

A doctor with good intentions could make an innocent mistake and tell a healthy individual that he has cancer. The fear factor alone could do irreparable damage to that person's health and well-being. In my opinion, a doctor should never give patient destructive information, such as predicted length of time he/she will survive. After such a prognosis from this authority figure, many patients lose all hope, and their bodies respond according to the predictions and begin to shut down. There are even well-documented examples of patients who were mistakenly diagnosed with cancer and who began to show visible symptoms until such time that it was discovered that their test results had been mixed up, and in fact, they never had the disease. A person's belief and the words of such an authority figure have tremendous power over their outcome.

Current Approaches to Cancer Treatment

Typical methods used to treat cancer, including surgery, chemotherapy, or radiotherapy will rarely be effective in the long term. In fact, if they are effective, it's probably more coincidence than an actual benefit of the treatment. On the contrary, these treatment approaches generally worsen the patient's condition, making bioenergy healing more difficult and sometimes impossible. Chemotherapy essentially poisons the entire body, weakening it, bringing the immune system to a zero level, and reducing its ability to recover from the resultant side effects.

Even more devastating are the side effects of radiotherapy. They can be compared to the following analogy: compare a tumor in a woman's breast with a tree with a damaged branch. The tree is sitting in the midst of a beautiful, very large green forest. It may seem reasonable to remove the damaged branch or the tumor so that new growth can occur. However, imagine that to accomplish that, a powerful flamethrower is used to burn off this inconsequential little branch. Yes, it will be successful at removing the damaged branch, but at what cost? In the process, it has destroyed so much more—perhaps the entire section of the tree, such that now, nourishment can no longer even reach that limb. The damage may have even extended to damaging the entire tree and a substantial portion of the forest surrounding it. The tree can no longer strengthen and restore that diseased limb as the flow of energy has been halted by essentially killing those sections that feed the tree.

Obstructions of Energy Lead to Disease

Consider a different model for the formation and treatment of cancer. A cancerous tumor may appear wherever the body's energy channels close or become obstructed. The precise reason for this obstruction may be quite different from one patient to another, but the underlying principle for the formation of the tumor is the same. Once the energetic and lymphatic channels become blocked, depriving that section of the body of its critical energy supply, a mass of negative energy accumulates. Consequently, this leads to tumor formation.

What causes an energy channel to become obstructed? There are potentially many reasons including emotional stress, a physical trauma or surgery, or even a severe blow, which can compress and ultimately block movement within the biofield. These are just a few of the possibilities.

The likelihood of getting cancer is frequently discussed in terms of both genetics and environmental factors. Scientific evidence suggests that our genes may predispose us to certain types of cancer. However, just because a parent, grandparent, or sibling was diagnosed with cancer doesn't mean we will necessarily have the same fate. Environmental factors must be present to trigger the cancer process. A group predisposition may also stem from the fact that several people lived in a similar environment. They may have shared lifestyles, hobbies and habits, diet, and stresses. They may have shared the same polluted air and thus contracted the same illness. Their common fate can also be explained from the state of the biofield or aura that surrounds them. An individual is surrounded by his personal energy field or aura, just as the members of his family, clan, or tribe are also surrounded by another commonly connected aura. This would explain how a group of individuals with some common bond may happen to contract similar cancers.

The aura is a representation of the energetic state of the individual. Consider this metaphor for the flow of energy within the body: imagine that your body contains a series of complex channels or pipes, analogous to the plumbing system supplying your home, except that these pipes are made of a flexible, elastic material. If sludge or other debris were to find its way into this "plumbing system," the blocked segment will start to swell, stretch, and enlarge in an effort to compensate for the increased pressure.

The blockage causes the energy channel to close down, creating a buildup of negative energy, which results in the formation of a tumor. In this model, which is quite different from the flamethrower example in the forest, to remove this harmful mass, it is only necessary to "flush out the contents of the pipe" in order to eliminate the blockage. Any unwanted debris will simply be flushed out with water. This is analogous to how bioenergy therapy works. This swelling resembles the development of a tumor in the body. If we can clean the elastic pipe and remove the unwanted material, the pipe would retract to its original form, and our flow of water would return to normal. This is essentially the same process that occurs during the early stages of tumor development. At this early stage, the tumor is an accumulation of negative, stagnated energy and has not yet begun to significantly alter the nearby organs or tissue. During this stage, bioenergy treatment is highly effective because it allows the channel to be cleaned, the blockage to be cleared, and the tumor eliminated—with no side effects, of course!

Why Surgical Intervention Is Frequently Counterproductive

A medical doctor will likely opt for surgical intervention. What is the impact of surgical procedures? By cutting out a section of the body's "energetic plumbing system" to remove the obstruction, not only is a section of the system now missing, but also often the nearby tubes and flow are affected as well. So instead of just eliminating the obstruction, other plumbing and energy-supplying systems in same area may be damaged or destroyed. This is why, in so many cases, new tumors appear after the surgery because the natural self-cleansing function of the lymphatic system has been weakened. When metastasis occurs in our body, this also can be explained using the plumbing system metaphor. If an obstruction in a major pipe is not cleared out and persists, it will continue to accumulate waste. This waste will spread, reaching up the pipe until it comes to a fork in the network of pipes from which it will continue to spread.

Usually once a tumor is present, the body's immune system will respond with its line of defense. If the body cannot completely eliminate the unwanted material, it will wrap it in a defensive layer to protect or insulate the surrounding environment. An example of this was my wife Elada's breast tumor. As you remember, the doctors discovered a cancerous tumor in her left breast and recommended surgery to remove it. We decided not to have the surgery, and she lived with it more than eight years before I removed it completely using bioenergetic treatment. Somehow in the interim, her body had found a way to supply the affected part of the body with the necessary life energy to allow her to live without any significant problems. In many cases, the body's energy supply is often only partially disrupted when the tumor encroaches on the energy channel. In this situation, the tumors are generally benign and can remain of little consequence for years. However, they can become malignant when this "insulation" is damaged and growth starts. When a tumor increases in size, to such an extent that it totally overlaps the energetic channels, a slow decay of the physical body can begin. At this point, we must find a way to remove the tumor.

Treatment Using Bioenergy

Early in my life I was led to believe that cancer was not only incurable but always spreading out of control. The very word—the C word—seemed to instill fear in everyone's mind. It was quite common, then, to think that reversing a cancerous process was a useless endeavor. Even now humanity believes that they can only dream about a cure. And from what everyone is led to believe, if that cure is ever to come, it could only be possible from the billions of dollars being spent at our high-tech scientific research centers. In the meantime, one of the oldest and most widely practiced healing systems on our planet is being largely ignored.

The practice of bioenergy therapy demonstrates that it is possible to control cancer. Cancer can manifest itself very rapidly in the body and vanish with the same speed. Our body is eliminating cancer every day. In order to do so, the immune system must be strong, fully charged, and properly "programmed"—and we must support rather than try to alter that "programming."

Unbelievable as it may seem, cancer is actually easier to cure than even any type of muscle pain. This is possible because cancer develops essentially as a result of a bioenergy problem and, therefore, can best be treated with the help of bioenergy-based interventional healing. I came to this confident conclusion after more than thirty years of the practice of bioenergy healing. During this period, I facilitated a cure in many patients who presented with cancer as well as a variety of disorders. Sixty-five percent of cancer patients are healed after the first course of treatment. (A course is generally ten to twelve consecutive days, followed by break of several weeks.) Another 20 percent see their cancers disappear after the second or third course of treatment. Based upon my experience, I have observed that cancer patients have a 90-95 percent chance of recovery. Because cancer is a bioenergy problem versus a physical problem, bioenergy treatment directly and profoundly affects the cancerous tumor, softening it, and gradually shrinking it away from the very start of treatment.

There is a fundamental difference between treatment using bioenergy as opposed to the administration of chemotherapy and radiotherapy. Chemotherapy and radiotherapy destroy healthy cells and damage healthy organs along with the lymphatic system in the process. Bioenergy, on the other hand, cleanses the body's systems and improves their functioning. It opens the energy channels and fills the body with pure and fresh energy, which gives the patient new vitality and the ability to rise above the illness. Scientific institutions worldwide are spending huge sums of money to develop new methods of fighting cancer. They enhance the technology of their radiotherapy facilities, find newer, stronger, and more poisonous chemotherapy drugs, but the results are and will always continue to be marginal and temporary.

A New Approach to Healing

The wrong approach to any problem will never yield the desired results. Although not very well advertised, this is what is being found with many mainstream cancer treatments today. It is simply impossible and illogical to cure an illness by destroying the body's immune system, which is its primary resistance to disease. This is what occurs with chemo and radiation therapy. On the contrary, it is necessary to strengthen the immune system to such an extent that it will be able to resist and overcome the disease. The most elementary difference between modern medicine and bioenergy therapy is in its approach to the body system.

Part of the reason for this complete divergence of approach is that modern medicine's recognition of what it means to be a human being comes from a limited and unfortunately ignorant perspective. Modern medicine only addresses the individual's physical body, which exists as a result of physical and biochemical processes in the internal organs. Bioenergy therapy considers all four aspects or "bodies" of a human being: the *physical, ethereal, astral,* and *mental.* Modern medicine also doesn't seem to recognize the degree to which our body has the intelligence to know how best to heal, if we provide it the support it needs.

The body given to us by our creator can be thought of as a unique, brilliantly designed, and organized "machine" driven by a powerful "minicomputer"—our brain. We can think of this minicomputer coming equipped with an operating system program capable of analyzing and responding to the environment around us. In keeping with our computer analogy, along with this built-in feature, our brain can use "new, more sophisticated add-on software" available from the existing base of universal knowledge. For instance, it may acquire an ability to destroy a growing tumor. Yet even with all of this complexity, our body is actually much easier to maintain than we are led to believe. Many people believe our health is maintained through the intervention of doctors and other "authorities" on the optimal functioning of our bodies. However, in fact, the body is capable of self-control, balancing, and self-healing when we give it the opportunity. But as with any "machine," to use it properly and to its full potential, we need to understand its inner workings. Unfortunately, this machine does not come with such an instruction manual.

The Four Components of the Human Body

As we have been discussing, the human body is not a singular entity. It consists of four interconnected bodies—the physical, the ethereal, the astral, and the mental. Let me briefly describe them to you:

1. **The Physical Body**
   The physical body represents our body as we know it—consisting of our flesh, bones, organ systems, etc. In order to keep our physical body in good health, it requires the following basic support:

   - sufficient oxygen, using the proper method of breathing
   - sufficient water to accomplish the various internal chemical processes and to generate the necessary electrical current inside the body
   - the required chemical elements provided by eating the appropriate foods
   - adequate physical exercise in order to maintain and improve strength and flexibility

2. **The Ethereal Body**
   The ethereal body represents our energetic body. It provides us with the necessary "life energy," which flows through our spine (sympathetic and parasympathetic nerves as sort of a main cable) and distributes itself into the life energy centers known as *chakras* from where it is routed to other destinations throughout the body.

3. **The Astral Body**
   The astral body represents the place from where our desires, both good and bad, originate. Stories of ghosts and hauntings refer to this body, which is the next station of the soul following death.

4. **The Mental Body**
   The mental body represents our spiritual body. It is the final harbor for the soul, where it will wait for the next incarnation.

All four of these bodies are intertwined and dependent upon one another. Weaknesses in one of these four bodies often have repercussions in one of the others. All of our illnesses usually start in the more delicate or higher (nonphysical) bodies, usually the astral body. In the case of cancer, it is an imbalance in the ethereal body and, therefore, can be corrected using cosmic bioenergy. But to be truly healthy we must care for all four bodies. This is the challenge not currently recognized by our current medical establishment.

Rudolf Steiner's book, *The Lord's Prayer: An Esoteric Study*, further explains how Jesus conveyed this knowledge to us of the existence of these four bodies as their symbolism is found in this commonly recited prayer. These four bodies are all a part of us, and as mentioned in my cases about coma patients, the existence of these "other bodies" explain how it is possible for patients to recover when their physical body and brain activity would suggest otherwise.

Simple Reminders about Your Healing Capacity

Remember as we have shown, diseases like cancer are an energetic problem of the ethereal body. Your body can be restored to health and balance through the infusion of bioenergy.

I am personally not in favor of many pharmaceutical drugs as the side effects are often little known and lead to other diseases. I had a firsthand experience with my father dying at an early age from the side effects of medications given to him for what began as a simple cold. The side effects created complications and resulted in more drugs being prescribed—one drug after another until it was too much for his body to endure.

Did you know that "being happy" is an extremely powerful and "nontoxic drug" that signals the initiation of many health-promoting processes within the body systems?

I believe enjoyment of life's pleasures is of the utmost importance, and as mentioned before, that includes indulging in some of your favorite foods—balanced with some of your newfound lifestyle and bioenergetic tools to keep your body temple well!

The immune system is the only "drug" needed to cure cancer. Dismiss the idea that cancer is incurable. If there is a tumor present in someone's body, it can be dissolved by meditation or by remotely sending bioenergy using the mind to concentrate on that area within the individual.

As an example, I personally believe that women should do a breast self-examination followed by bioenergy. I am concerned about their exposure to the radiation of a mammogram, which is already a known risk factor for cancer. If there is some questionable hardness discovered tactilely, it can be dissolved through meditation. Using two or three fingers over the lump, imagine that the abnormally thickened or hardened tissue is a piece of ice and you are melting it using the energy from your fingertips. You will be surprised how quickly a tumor will soften and shrink. This may happen within a few minutes, a few days, and of course sometimes longer. There is no one consistent formula. Even with all my years of experience in this technique, for some people, the response doesn't appear obvious until after a second course of therapy. And that may have begun as many as ten days after the completion of the first course. Every case is different, yet most do respond to bioenergy if we are optimistic, patient, and persistent.

Knowing that even one of the miraculous stories I have described about individuals who were healed from cancer is possible can be the encouragement that seeds the brain and the belief system of someone facing a similar challenge. That stimulus can "program" the brain to command and turn the immune system into a powerful force and restore the body to a healthy state.

Epilogue

As you know, I eliminated the breast cancer in my lovely wife, Elada, ignoring the warnings about the possible dangers of working with cancer using bioenergy. Over the years, Elada has been a keen, quiet observer as I studied various healing modalities. She was often present in the background as I taught many a student hypnosis, yoga, and bioenergy techniques. Although not officially part of "the class," with her repeated exposure to the techniques, they were fortunately ingrained in her subconscious and would come to save my life.

On February 1, 1997, Elada and I were in a serious car accident. Our younger daughter Arevik's godfather, Yuri Gilavyan, was visiting from Armenia. So we wanted let him do the touristy thing and took him to Las Vegas.

My wife loved to drive, so we decided to take her new 1997 Toyota, which we had just purchased a month prior. Oddly, when we had taken the car to our insurance agent and longtime friend to obtain our insurance paperwork, take the photos, etc., he advised us to change all four tires. I found this unusual as there were minimal miles on the car. He indicated that the car was equipped with a brand of tires that had been the subject of numerous court cases since 1993. He cited fifty-eight accident cases and four deaths that were attributable to these tires. Elada and I discussed it and thought that perhaps we should just monitor them more closely for excessive wear. It seemed unreasonable to replace what appeared to be perfectly good tires. So we began our trip to Las Vegas in our new car. Everything seemed fine, and I made a point to check the tires when we stopped for fuel and lunch as the insurance agent's concerns were still etched in my mind. The tires showed no unusual wear, so we continued on our journey.

Yuri was in the passenger seat and I was in the backseat. After only ten minutes back into our trip, suddenly there was a loud noise from the left rear of the vehicle. It was the tire. It had exploded! Elada tried her best to keep the car on the road; however, finally it was too much for her to handle, and it veered to the right and ran off the highway into a vineyard. I hit my head and lost consciousness.

Paramedics rushed us to Barstow Hospital. Elada suffered a broken clavicle. Yuri had multiple fractures in his left arm and leg. They immediately operated to insert metal pins in order to stabilize his bones so that they could heal cleanly. I was not as fortunate. My seat belt did not activate to restrain me, and I suffered a major head injury with six broken bones in my face and twenty-three shattered teeth. Nineteen of my ribs were broken, many of them from both sides. Two of the broken ribs had impaled my lungs, which were now completely full of blood. With all of this trauma, I had slipped into a coma.

The doctors began to repair some of my broken bones but believed my condition to be hopeless. They advised Elada that if somehow I beat the incredible odds and awoke from this coma, I would be a vegetable. They strongly recommended that the life support systems be withdrawn.

Elada's repeated mantric response to them was "Doctor . . . does he have a heartbeat?" They would respond to Elada patronizingly, "Yes, but, Mrs. Manukyan, you must understand that he . . ."

Needless to say, Elada ignored their grim counsel and continued to hold her ground, insisting that I be left to recover. There was to be no removal of the life support. She was taking matters into her hands.

Elada decided that she would attempt to stimulate the healing process in me as she had seen me demonstrate on my patients time and time again. Although I had never formally instructed Elada on bioenergy therapy, she was a highly resourceful woman and knew she was determined to find a way to use these same techniques to bring me back to our world. She moved in to a nearby motel, staying with me day and night, long after hospital visiting hours. Knowing that on some level I could receive her communications, she talked to me continuously to maintain the bridge to my subconscious. Elada recalled what I had said about my earlier coma patient, Jennifer. I had explained to Elada that while the conscious mind was in a very deep sleeplike state, the subconscious was in fact alert and able to receive and process verbal directives. The combination of bioenergy and verbal directives was how I was able to bring Jennifer back to a conscious state. Many modern medical doctors are unaware of this other level of heightened awareness found in a comatose patient. Elada knew that I could hear her on this level and that I knew instinctively how to heal myself. So she began to use a similar approach and instruct me as to what areas were in need of healing and asked me to assist her in sending bioenergy to those areas.

The medical staff continued as not very optimistic at all about my prognosis and would not have understood what Elada was doing. In fact, one situation that actually resulted in somewhat of a chuckle even under such grave circumstances was when one of the nurses saw Elada for the first time waving her hands across my body in a gentle ceremonious manner. Having no ideas that she was directing bioenergy, the nurse immediately assumed the worst as though Elada was saying her good-byes. When in fact she was bringing me more strength and healing in every moment of her heroic efforts. Slowly, through Elada's persistent daily work with me, we began to gradually pull me out of the coma. Finally, forty-six days later, I awoke from the coma and once again amazed another group of astounded physicians with a "miraculous and impossible" recovery.

So as you can see, Elada too had within her the capacity to perform bioenergy healing as we all do. Under those grave circumstances, she was able to muster up the ability to reach into her memory banks and go within her soul to find the inner knowledge she needed to connect with me and my innate healing abilities.

Postcoma Insights

After awakening from the coma, I recalled and recounted many experiences during that "unconscious" time with remarkable accuracy. One very clear memory I had was that my mother-in-law and I were traveling to many destinations together that I had always wanted to visit. It seemed so incredibly vivid and real. Elada's mother appeared young and fit during these virtual "trips" together. When I awoke, I immediately suggested to Elada that we apply for a U.S. visitor's visa, so her mother could spend some time with us. Unbeknown to me at the time, Elada appeased me for that moment as in fact her mother had passed away in 1992. She did not want to remind me of that fact, which might explain how her mother was able to "visit" me in my comatose subconscious state.

I had received another amazing "visitor" before I emerged from the coma. I told Elada that my father's cousin Armik had appeared to me. Arnik was in charge of the Armenian Republic Hospital, and while in Armenia, he had sent high-ranking officials to me requesting healings. Elada was skeptical about my claim that I had seen him "on the other side" as she believed that was impossible as he was still alive . . . or so she thought. Just days later, we received call from his son in Armenia who told us that his father had in fact passed away just days ago.

A third "transient companion" on the other side that for me validated the existence of other worlds or dimensions was a man named Alexander. Alexander was the father of Elada's close friend, Anahit. Anahit, Elada, and I had all worked together in the Yerevan Drama Theater many years ago. Anahit was one of Elada's best friends, and her husband and I were also good friends for over twenty years. After the car accident, naturally Anahit had made a regular presence at the hospital to support my wife during my long recovery.

Sometime later, after I awoke and returned home, I had the opportunity to share with Anahit more details about my experiences while in a coma. I told her that her father Alexander had come to me during that time to keep me company and assured me that I would recover and go back to my family that needed me. Anahit was in awe and in disbelief, not so much because I claimed to have seen someone while in a coma, but because she knew I had never met her father nor seen a photo or any other likeness of him. I explained to her that the "mystery man" claimed to be her father. So the next day she returned with a photo of a group of people and asked if I could identify her father. At a quick glance, I was able to immediately point to his image. She was astonished as he had passed away last year, and we had never met. After seeing this evidence of a connection to another dimension, Anahit revealed a dream she had had the night before our accident.

Here is her description of the dream and the surrounding events:

My husband was up into the wee hours as always—a computer technician, working in the midst of his disparate spare parts inventory in the living room. I finally turned in for the evening and fell asleep immediately. My dream began with me standing in the middle of our living room; then, a moment later, I found myself in front of the basement door, which was closed. There was a continuous monotone sound coming from behind the door. I cautiously opened the door to investigate the noise and drew back in fear as four or five spirit forms flew through the doorway and into the living room, encircling me. One of them stopped and stood in front of me. I realized with some sense of relief that it was my father. He greeted me and then informed me that the following day I would receive some distressing news—that one of my closest friends would be involved in a serious accident. He continued by telling me that the attending physicians will be dismal about my friend's prognosis and potential survival. Upon hearing this alarming news from "my father," I became distraught; however, he calmed me and assured me that the friend would in fact pull through after a period of time. He then quickly vanished.

Upon waking from the dream I found myself in a strange and uncomfortable mood that I couldn't seem to shake. Soon after, I received a telephone call and the shocking news of Anush and Elada's accident. My husband and I left immediately for San Bernardino. We knew our presence was the best we could do for them at that time and must have been the reason for my father's warning.

Anahit did remain closely by my wife's side in San Bernardino with continuous words of encouragement and support. With no significant progress, the staff at the hospital recommended that my wife transfer me to the general hospital in Los Angeles. It was better equipped and, in their opinion, afforded someone in my condition the only hope of recovery. That hospital was also very close to our home in Los Angeles. Overall it would make for a better arrangement so that my wife could get home more frequently to tend to our children. Taking all these factors into consideration, my wife managed to get me transferred to Los Angeles after seventeen days of being in the San Bernardino hospital.

Finally, after her continued efforts to communicate with my healing instincts and keeping the doctors coercions to discontinue life support at bay, on the forty-sixth day I awoke to the amazement of my physicians. I started my slow and incremental recovery. Even after I came out of the coma, the doctors still didn't have any words of encouragement for my wife. In their opinions, it was a miracle I survived the accident and a second miracle that I came out of the coma. They continued to hold a grim prognosis for me of life in a wheelchair as a vegetable, and as a result, I was given a label of "disabled" by the government. I decided to prove to them that there is no shortage of miracles. I began the recovery process of my body, mind, and spirit and was even practicing yoga at my previous level of proficiency in just one month. Thank God I still retained the ability to perform bioenergy healing as that part of my memory was intact. In some ways, my healing abilities were enhanced;

however, because of my energy level, I no longer could heal fifteen to twenty people each day as part of an advertised healing practice. Instead I might work on two to three people occasionally. They were often friends and relatives in need. There were still many people over the years with serious challenges that would request and receive miraculous healings. You too can learn to heal using bioenergy healing, just as my lovely Elada did to save my life. She has since continued to assist me with some of the patients, and she too has become a very powerful healer.

So once again it is possible for *all of us* to learn to use bioenergy—including you!

Appendix A

Creating a Healthy State Using the Power of the Bioenergy Field

I have been involved with the practice of bioenergy therapy for twenty-seven years. Leading up to that, I've described the many healing modalities that I have studied, mastered, and incorporated into my work along with some of my unique techniques. As I partnered with my patients to conquer these diseases, some of which were considered incurable, the veritable simplicity and limitless possibilities of bioenergy healing never ceased to amaze us all! I only wish that there were some way for everyone to know and be helped!

I have a great desire to share my skills and experience so that the ability to control and work with bioenergy will be available to as many people as possible. My first thought about writing a book came about ten years ago. I really wanted this to be different from other medical books and interventions and eliminate the notion of the "incurable." My goal was to document all of my experience in natural healing using bioenergy, also known in some philosophies or cultures as spiritual energy or *prana*. When we see through the many compelling patient cases how remarkable our body's healing abilities are in response to an infusion of bioenergy, we know anything is possible.

Later in this appendix, I will provide some techniques to begin practicing and harnessing bioenergy to promote healing. You too can carry the tools and develop the abilities to master your bioenergy healing skills.

At the same time, I recommend that you learn how to protect and maintain your health through yoga and other modalities that will support your body's life-energy capacity and flow. I would like for the teachings in this appendix to be the catalyst for that process—to be a guide and fresh "food for thought" for you, as along with our doctors, chiropractors, extrasensory healers, massage specialists, and healers of all kinds. We all wish to help one another to be healthy and experience full enjoyment of our lives.

The remaining portion of this section will help you to develop your skills with the following series of exercises.

EXERCISES

Overview

With the principles and exercises that follow, you will gain an awareness of your body's energy system, and the ability to sense and control it. Eventually you will be able to pinpoint organs that are out of balance, diseased, or simply not functioning optimally. By working with bioenergy, you can learn how to maximize the functioning of your organs, eliminate digestive disturbances, relieve headaches, muscle pain, and even address serious diseases and virtually anything out of balance! You can learn how to clean the lymphatic system and strengthen your immune system. If you or any other individual you are working with has already had medical interventions, such as surgery, chemotherapy, or radiotherapy, the bioenergetic techniques will help to overcome the associated side effects and promote the course of recovery.

With practice, your ability to sense and manipulate the biofield will all become second nature. Your skills will develop both with an individual at close proximity or even with someone at a distance using telepathy. Occasionally a photograph can be used as an aid as you perfect the technique of remote healing.

To begin your training, we will exercise and strengthen your ability to concentrate and exert willpower or a strong intention on a single point of focus. This is an important foundational skill in learning to manipulate bioenergy and create a healing effect.

Concentration and Willpower Training

It takes some discipline and practice to focus on one thought or one body part for a long period of time. Using these exercises, you acquire this technique of removing the clutter of surrounding distractions. This next exercise will help you to improve your concentration, your willpower, and an "extra bonus"—your eyesight! I believe that it is essential that one develop both strong willpower and some level of hypnosis ability.

Exercise: Fixed Gaze

1. Begin by taking a blank sheet of standard white paper (about 8 1/2 by 11 inches).
2. Draw or paste a dark solid black circle of about one inch in diameter in the center of the paper.
3. Attach the sheet of paper with the black dot/circle on the wall.
4. Ensure that from a seated position, it would be at your eye level.
5. Sit at a distance of about 6-6.5 feet from this black dot and focus your eyes directly on the dot.
6. Try not to think about anything else—just focus on the dot.
7. Your task is to focus your eyes on the black dot without blinking. If you blink, relax, and begin the exercise again.
8. Each time you attempt the exercise, try to increase the period of time that you can remain focused without blinking.
9. The goal is to eventually remain focused on the black dot without blinking for as long as twelve to fifteen minutes.

More Willpower Training

This exercise, like the previous one, will strengthen your willpower and your eyes and enable you to withstand any penetrating looks directed at you. When someone stares directly at you, this will give you the ability to resist yielding to another individual's willpower and maintain your gaze forward without lowering your eyes. Remember that the person who lowers his/her eyes first is considered the one with the weaker willpower. I believe this is very important to know as we communicate with the surrounding world every day. If you are engaged in a business discussion, avoid looking into the other person's eyes when he wants to influence you to accept certain conditions that you may not be in favor of. Of course, if you want him to accept your conditions, make direct eye contact with him at the right moment.

Exercise: Mirror Training

1. Sit comfortably in front of a mirror at a distance of about 3-3.5 feet.
2. Look directly at the point in the mirror where your eyebrows meet the bridge of your nose.
3. Remain focused in this position as long as possible while holding your eyes open without blinking.
4. Note the period of time you were able to successfully maintain the exercise while trying to reach a new record each time.

PERCEIVING YOUR BIOFIELD

The following exercises will help you to learn to perceive and ultimately manipulate your biofield. **Preparation—Pharaoh Position**

1. Begin by assuming the *pharaoh* sitting position. To do this, sit upright in a straight-backed chair, totally relax your body, and concentrate on just two points: the top of your head and your feet.
2. Visualize and allow the stream of bioenergy to pass through you, beginning at the top of your head and continuing to your feet.

This subset of exercises will help you to strengthen the perception of your palms. If you do these exercises every day, you will soon be able to feel your biofield and the biofield of others.

Exercise: Sensing the Biofield with Your Palms and Fingertips

This exercise allows you to perceive your biofield. It demonstrates that you have control over the field and that you can sense its power.

While remaining conscious of this energy stream you have tapped into through your preparation, do the following:

1. Raise both arms directly in front of you, parallel to the floor, with the palms of your hands nine to twelve inches apart and facing each other.
2. If the energy stream is strong, you may sense a slight heaviness or a prickly feeling in your palms.

3. Begin to move your palms toward each other very slowly, bringing them farther away again. You may feel the biomagnetic field *between your palms*, which will compress as your hands come closer and will expand as they move apart.
4. You may feel as though you are moving two magnets toward each other and then away from one another.
5. Now leave your left hand extended for a moment without moving it and move your right hand up to the point of the curve of the left *palm* and then down to the tips of the fingers.
6. You may feel some movement in the left hand as if it were a gentle breeze. If so, you have experienced the biomagnetic forces of the right palm, which intersect and go through the left palm.
7. Now change hands and repeat the same exercise.
8. Next, we will use our fingertips for finer, more precise "laser beam" control and sensing of the energy field.
9. Hold the left arm still and direct the *fingertips* of your right hand toward your left palm.
10. Start moving your *fingertips* down, forward, and backward. Make very small circular movements, both clockwise and counterclockwise.
11. You will feel the same synchronous movement of the biofield in the palm of your left hand.
12. Now change hands and do the same exercise.
13. Upon completing this exercise, place your hands on your knees with your palms up, and relax completely for a moment.
14. Then repeat the exercises we just discussed.
15. Once you sense the movement of the biomagnetic field in your hands, you are ready to move on the next exercise.

Strengthening the Sensitivity of Hands

Exercise: Sensing Objects in the Biofield (using one hand)

Sit at a clear table in the *pharaoh* position. While letting the energy stream pass through you, do the following:

1. Raise your arms, keeping them parallel, with palms facing each other.

2. Now, begin moving them slowly, bringing the palms closer, then apart.
3. You will start to feel the biomagnetic field between your palms widening when you increase the distance and compressing when you decrease it.
4. Now turn the palm of *your left hand* down so that it faces the tabletop, at a distance of about three to four inches above the table.
5. Moving your palm slowly up and down, try to feel the bioenergetic field formed between your palm and the tabletop.
6. Since there is nothing on the table, this field will be smooth and level.
7. Now, place a small object on the table, such as a pen or pencil.
8. Again, move your palm over the table.
9. When your palm passes over the object, you should sense the field density change. It will no longer be uniform. You will have the feeling of crossing a barrier.
10. Now, substitute another object for the pen, and try again to feel the change in the biofield that corresponds to the presence of the object.
11. Practice this exercise frequently.

Additional Note: Plants have an easily detectable biofield and are, therefore, very valuable for novice practice. They are quite sensitive and will actually return your bioimpulses. You can detect plants using your palms and take note of the differences in sensation between plants with healthy, well-hydrated leaves and those that are drier and crumbly.

Exercise: Sensing Objects in the Biofield (using both hands)

Now repeat the exercise just described; however, use both hands and different parts of each hand to sense the objects. Concentrate on feeling the objects with your fingers or with the sides of your palms.

Exercise: Creating a Line of Energy

This exercise is particularly useful for our healing purposes. It will be particularly helpful when we are evaluating smaller, more contained areas in the body.

1. This exercise is practiced outdoors near a tree.
2. Stand straight, in proximity of the tree, with your hands at your sides.
3. Relax and allow the energy stream to flow through you.
4. Mentally, get in contact with and "connect with" this tree.
5. Now create an "energy line," that is, a direct energy connection between you and the tree.
6. To do this, you'll have to concentrate on the tip of your index finger, let's say, of your right hand, until you feel the heaviness of the energy stream.
7. Now imagine that you have a very small ball of energy on the tip of your finger. With one movement of your finger, throw this ball in the direction of the tree and actually direct it to the tree.
8. Once it reaches the tree, bring it back, taking it into the palm of the same hand. Of course, you will have to see all of this action mentally, in your imagination.
9. Imagine throwing the ball and having it fly directly to the tree, leaving in its wake an energy trail. It's as though you have mentally created your laser beam, which will remain as long as you wish.
10. Now hold that feeling and wait until somebody or something passes through that line. Interestingly, you will actually feel the movement of that person or object in your hand.
11. Without moving your head or your hand, you will be able to detect how many people or objects have intersected your line.

Programming the Biofield

Just as we know that a simple electrical current, flowing through a thin cable, can deliver programmable pictures, sound, music, or fax data into our homes, the bioenergy stream passing through our bodies contains certain programmable data, which can be decoded, corrected, removed, and reprogrammed to suit our needs. We can use this energy to influence our body's immune system and to heal compromised areas. To do so, we must be able to do much more than just feel the energy stream. We must also learn to program the incoming energy in order to produce the desired effect.

Exercise: Creating Cosmic Energy (Bioenergy) Balls

1. Sit in the *pharaoh* position and let the energy stream go through you.
2. Raise your hands parallel to each other, with your palms facing each other.
3. Now, slowly move your palms, first toward each other and then away, until you can feel the biofield between your palms becoming dense and heavy.
4. Keeping your hands still for a moment, imagine that you have a one inch in diameter golden ball in each hand. Concentrate on these imaginary balls and feel them in your palms.
5. Fill them with cosmic energy, flowing directly from you through your palms.
6. Now, imagine that the balls begin to emanate cosmic energy itself. It feels like a thousand small needles penetrating your palms.
7. Next, control the energy with the power of your thought. Switch it on, or switch it off. You control it.
8. Take a break, relax, and perform this exercise again repeatedly.

Exercise: Moving the Energy Ball

1. Allowing the ball to remain between your palms; let it continue to emit its energy.
2. Hold this position for a few minutes and then, using only the power of your mind, try to elevate the ball slowly from its initial position. Do this very slowly, focusing your concentration entirely on the ball.
3. Once it has risen three to four inches, begin to move it down again slowly to its original position.
4. Feel the energy stream with your palms as it passes through you.
5. Stop the emission of energy from the ball and then commence again.
6. Repeat this exercise several times.

Exercise: Working with Multiple Energy Balls

Sometimes, because of physical, emotional, or spiritual stress, we may suffer a significant leakage of energy from a chakra. If such a condition is not addressed, neighboring chakras may also be weakened, which can subsequently impact body organs. The following exercise will prepare you to remedy this situation.

Part 1

1. Sit in the *pharaoh* position and let the energy flow through you.
2. Raise your hands parallel to each other, with your palms facing each other.
3. Slowly move your palms toward each other and then away, until you can feel the biofield.
4. Again, imagine that you have the small energy-emitting ball between your palms. But this time, imagine also a second ball just a few inches above the first.
5. Focus on the energy emanating from both of them.

Part 2

6. Now imagine that the first ball is the sun—albeit a very small sun. It is the source of all energy. The second ball will be the receiver.
7. This time, focusing your concentration, direct the first ball to emit a fine stream of energy to the second one—completely filling it.
8. Once this is accomplished, reverse the procedure and cause the energy to flow back into the first ball.
9. Then take a break, relax, and prepare for the next part of the exercise.

Part 3

10. In this part of the exercise, instead of causing the first ball to *send its energy to* the second, focus your concentration on causing the second ball *to take the energy from the* first.

11. You will see the same fine stream of energy between the balls, but now you must feel the energy being drawn from the first ball to the second rather than being sent.
12. The action is quite different. Now reverse the process again, and cause the first ball to take the energy back from the second.
13. Then relax for a few moments before continuing.

Part 4

14. Finally, imagine the same two balls in play as before, one being the sun—the source of energy—the second emitting no energy at all.
15. At this point, mentally place a third ball in such a way that the energy-emitting ball is between the other two.
16. Create your exercises filling the second and third balls from the energy of the first and then reversing the process and causing the energy to be drawn back into the first.

If your imagination is good, you can improve and create a multitude of exercise scenarios.

Eventually this technique may be used effectively in balancing and cleaning the chakras, which become interchangeable with the energy balls.

Controlling and Programming Your Biofield

The following exercises will help you to control and program your biofield.

Exercise: Flower and Seed Growth Visualization

Preparation:

1. Sit in the *pharaoh* position and concentrate on the energy stream flowing through your body.
2. Elevate your arms and hands so that your palms are again facing each other.

3. Slowly bring them toward each other, and then away, until you can feel, or sense, the biofield between your palms.

Part 1

4. Mentally place a flower—a rose, for instance—between your palms and imagine that the petals of the flower are closed.
5. By using your power over the biofield, open the flower, petal by petal. Imagine that as soon as the rose is completely opened, it begins to glow—emitting a bright, powerful, golden stream of energy.
6. Now, after a few seconds, begin to close the petals one by one, until the flow of energy ceases, and the glow has disappeared.

Part 2

7. Now substitute the image of the flower with one of just seeds.
8. Slowly fill the seeds with a stream of your energy and observe how the seeds begin sprout and gradually send new shoots and leaves upward.
9. The leaves slowly become new flowers, and under your control, the flowers open fully.

Part 3

10. Finally, eliminate the image of the flower, and return to just your empty palms facing each other.
11. Slowly move your palms closer together, then apart, closer together, and then apart again.
12. Imagine you can feel the biofield between your palms becoming denser and then darker.
13. Now we will gradually replace this heavy dark energy until it becomes the bright golden glow of our beautiful flower.
14. Do this by concentrating on your control of this energy as you pour a powerful stream of fresh new energy into the dense, dark cloud.
15. Watch as it gradually changes to a bright, healthy, golden glow, and begins to emit its own fresh new energy.

Working with Energetic Centers or Chakras

Bioenergy is the primary power source for all natural life. It energizes and maintains the life-creating processes. It enters our bodies like a stream—an energetic column. Stop this flow of energy, and the body will die. Our spine acts as the main cable through which this life energy is transferred to the *chakras*—the little energy centers previously discussed in chapter 2. These chakras can be compared to small energy substations through which energy is distributed to the "consumers," our body's primary organs. The chakras also control many of the life-sustaining processes that take place within the organs. If the energy supply is disrupted, our organs function poorly. Healthy chakras will spin, constantly delivering energy to ensure well-being.

Chakras have an additional function of serving as windows into the spiritual world from which they acquire and decipher considerable information. The chakras are the center of our bioenergetic system and, in turn, the various levels of consciousness, ranging from bodily to spiritual.

Your body's bioenergetic capacity and flow will improve as you learn how to open and clean these chakras, which receive, accumulate, and distribute energy throughout the body. Healthy chakras assure the continuous flow of life energy to the physical body as well as the other three energy levels. Occasionally a physical or emotional problem may trigger the temporary malfunction of one or more chakras. For this reason, it is necessary that we regularly clean, open, and fill the chakras with bioenergy.

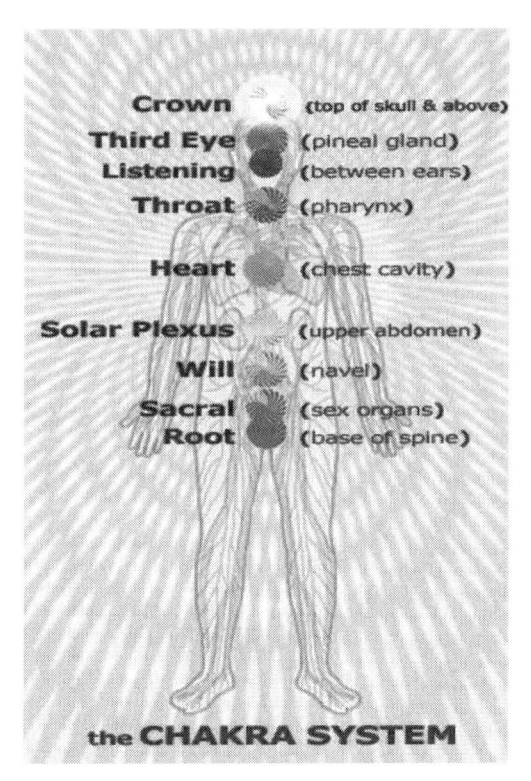

Figure 1. Nine Chakra System

As previously stated, we have nine chakras (see Figure 1. Nine Chakra System). Our "cleanest" bioenergy is found in the uppermost chakras. The "dirtiest" is in the first, or lowest, chakra (i.e., closer to the material than to the spiritual). Your job will be to keep the chakras clean beginning from the upper to the lower chakras.

Before starting the exercises for cleaning your chakra/energy centers, a few words are necessary regarding their actual locations in the body. Today in bookstores and libraries, you can find many books about chakras with detailed descriptions and illustrations. In all yoga books, you will find references to seven chakras. However, during the last two thousand years, humanity identified two additional chakras, which would not be found in any books for East Indian origin. You have to understand that the chakras are not a material entity and, thus, are practically impossible to locate with any precision. When "viewed" straight on, a chakra can be described as a clockwise-spinning, horizontal electromagnetic field. However, it's easier to identify a general area than a precise location. The approximate locations on the body are indicated in Figure 1.

Sensing Energy Flow and the Location of the Chakras

While working with chakras, don't concern yourself with the exact locations. It is preferable to simply concentrate on the level where the chakra works. Later, you will be able to feel some heaviness of the energy in the spot where it is located.

Sensing the Energy Flow

1. Assume the pose of *pharaoh* and allow the bioenergy stream to go through you.
2. Look straightforward without blinking if possible.
3. Concentrate completely on the biomagnetic stream going through your body for several minutes.
4. Mentally assess and test the flow of the energy through different parts of your body.
5. Wherever your perception determines that the stream is weak, focus and stay in that area longer.
6. When you have evaluated your entire body in this manner, proceed to the following exercises.

Sensing the Chakras

Part 1

7. Without losing the perception of the biofield, concentrate your mind on the approximate location of your first and highest chakra. In yoga, this is called the *Sahasrara*. We call it the chakra of *thinking* or the chakra of *wisdom*.
8. If you have a clean perception of this level, hold your concentration there for one or two minutes after which you may relax.
9. Repeat all this with the second chakra, third chakra, and so on.
10. When you have finished with all nine chakras, relax for a moment and then proceed to the next step.

Part 2

13. Again, assume the *pharaoh* position and concentrate on the stream of energy brilliant golden energy, the color of a halo, is passing through you.
14. Concentrate on the level of the first or the highest chakra. Imagine that a narrow horizontal beam of gold cosmic energy (bioenergy) is entering this level.
15. Allow the energy to fill this chakra level while maintaining your concentration on the chakra.
16. Once it is full and radiating energy outward, "lower" the beam of energy to the next chakra.
17. Fill the second chakra with this energy, feel it radiate outward, and then continue with the same procedure to the third chakra.
18. Always remember that you must sense that all preceding chakras are filled with cosmic energy simultaneously before moving to the next chakras.
19. Continue this exercise, filling each subsequent chakra until all nine chakras are filled with brilliant gold, cosmic energy.

Appendix B

Getting Well . . . Keeping Well

It is one thing to overcome an illness. It's another to stay healthy. My years of bioenergy healing have taught me how important this really is. It's not about cornering and eradicating every disease that comes your way; it's the prevention in the first place that gives us quality of life. Unfortunately, it seems that many of today's traditional medical doctors have not yet grasped this concept. Often cancer treatments are completed, and the patient is said to be "clear"; however, there is no consideration as to how they contracted the cancer in the first place and what will be done to prevent another occurrence.

It has always been my belief that a successful treatment of any sort is just the first step. Next the patient needs to know how to strengthen and care for his/her immune system and entire body in order to maintain their newfound health through their future.

When you own a car, you are taught the benefits of ensuring that you fill the tank with quality gasoline, change the oil frequently, flush the radiator fluid, and more. For some reason, we often know more about the maintenance of our cars than we do our own bodies. Do we know what foods are best for our body? Do we know how much water to drink? Do we know the important role of our lungs, and how to maximize their healthy function? To contribute to our knowledge, doctors must have a more holistic and comprehensive view of the patient and what creates a healthy state for their body and mind. In a perfect world, a physician would receive supplemental training beyond what is currently required in medical school. This would include an understanding of physics, a command of hypnotism, a mastery of yoga, proper breathing technique, therapeutic massage, nutrition, and certainly the capability to use *bioenergy therapy* for diagnosis and treatment! With this more comprehensive background, there are a number of foundational concepts that the doctor could share with their patients to ensure long-term health and well-being.

These include at a minimum an understanding of the following:

- effective bioenergy *prana*-enhancing breathing techniques
- the far-reaching role of water and its influence on all bodily systems
- various diet and nutritional guidelines appropriate for a specific individual's needs
- appropriate physical exercises from various schools, such as yoga
- the concept and significance of human energy meridians and pressure points

During my many years of practice, in addition to treating patients using bioenergy, I have emphasized the importance of their becoming completely self-sufficient in the care and maintenance of their own bodies and advocating the use of these concepts.

Again, I hope this section of the book will help you in your path.

Enhanced Breathing

We know our lungs and blood are the vehicle for oxygen to reach all parts of our body. To support and enhance the role of the lungs, I recommend learning and mastering yoga breathing exercises known as pranayamas. These practices teach us how to breathe slowly, deeply, and efficiently. The yogis said that God gave us a specific quantity of oxygen for our lifetime so we want our breath in the most effective manner to maximize longevity.

Breathing Experiment and Introduction to "Full Breathing"

Try this breathing experiment. From a relaxed seated position, count how many times you breathe in a minute naturally and record that number. Then make an active effort to breathe slowly and then count again your inhale/exhales per minute. Believe it or not, highly advanced yogis are so efficient in their breathing capacity that they can breathe only one or two times over the course of a minute!

Our lungs have three groups of muscles. One group of muscles controls the chest area, a second group controls the middle part of lungs or solar plexus area, and the third group covers the abdomen and controls the muscles in the area of the diaphragm.

In order to breathe effectively, you shall learn how to perform *full breathing*. This entails filling all three parts of the lungs that we just described: the upper/chest area, middle/solar plexus, and lower diaphragm/abdominal area. This will require developing muscle control and strength in all three areas. To do this, you will isolate each muscle group individually and perfect the breathing in that respective area. Begin with the lower abdominal area and fill that portion of the lungs. Once you feel it is full, continue to expand your breathing upward to the middle and then finally the upper chest part of the lungs. This allows you to use your complete lung capacity. To learn more details about "full breathing," consult a *hatha yoga* book and refer to the section on pranayama. We will also provide some *full breathing* exercises in this book under the pranayama section of *Appendix B, Getting Well . . . Keeping Well.*

Water—The Elixir of Life

Water's Molecular Structure Responds to Our Energy

Water is essential to life for human beings and to create all universal life. Although I completed rigorous graduate work and became a quantum physicist, I received very little training about the power of water. In fact, it is now known that water acts as an incredible computer of sorts through which unlimited quantities of information can be programmed and stored in its structure as "memory." With information stored within, the water can influence other living things. Believe it or not, studies have demonstrated that the molecular structure of water can change based on its surroundings. For example, various types of music, the intent of our conversations, our thoughts, and our meditations hold an energy that is received by any nearby water and change its structure. How is this possible? Water appears to be a "dead" inanimate object, when in fact, water responds and becomes everything from a life-giving elixir to a destructive force depending on the energy of the thought forms it receives. By simply talking in the presence of water, although it stays the same chemically (i.e., hydrogen and oxygen), its structure is altered. For example, these photos taken by Dr. Masuru Emoto are a compelling demonstration of this concept. They portray the condition of water molecules under the influence of various words being "displayed." You may see many examples of these experiments on Dr. Masuru Emoto's web site. These are just a selected few. He placed the following words on containers of water and the diverse effects on the molecular structure of the water appeared.

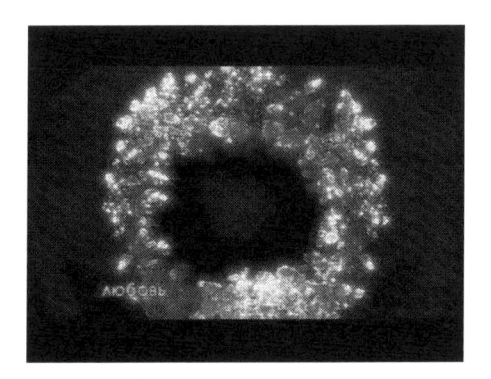

© Office Masuru Emoto, LLC
Water in the presence of the words *I Love You*

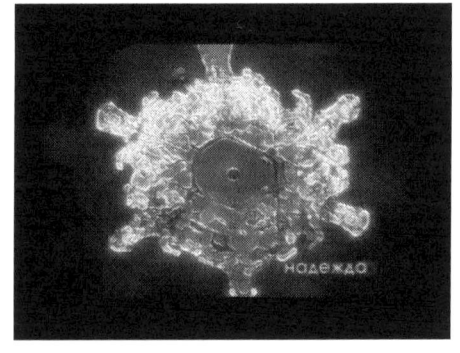

© Office Masuru Emoto, LLC
Water in the presence of the word *Hope*

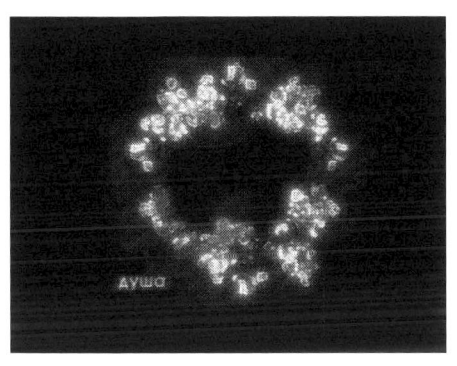

© Office Masuru Emoto, LLC
Water in the presence of the word *Soul*

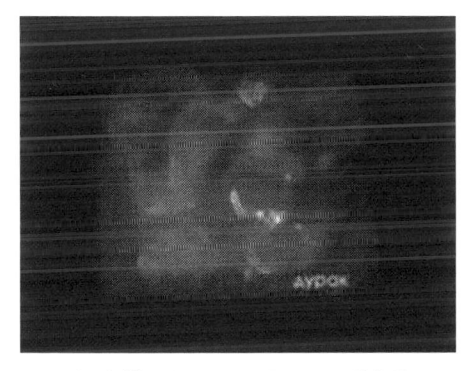

© Office Masuru Emoto, LLC
Water in the presence of the word *Fool*

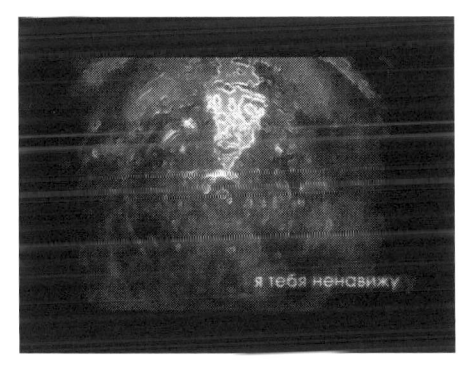

© Office Masuru Emoto, LLC
Water in the presence of the words "I hate you"

Getting Well... Keeping Well

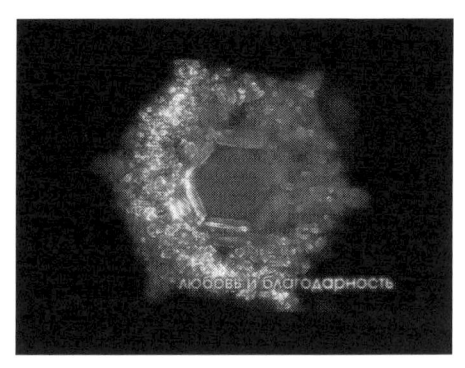

© Office Masuru Emoto, LLC
Water in the presence of the words *Love and Gratitude*

Water can change its "mood" or its molecular structure under the influence of various genres of music.

© Office Masuru Emoto, LLC
Water played the music, *Bach, Aria*

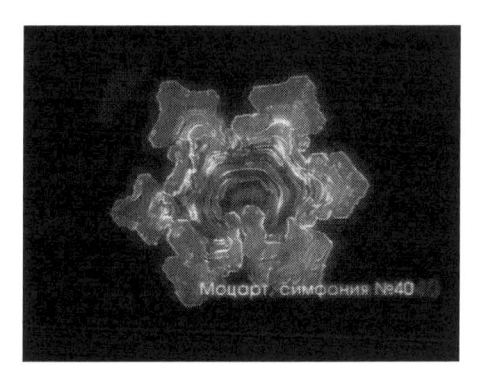

© Office Masuru Emoto, LLC
Water played the music, *Mozart's Symphony No. 40*

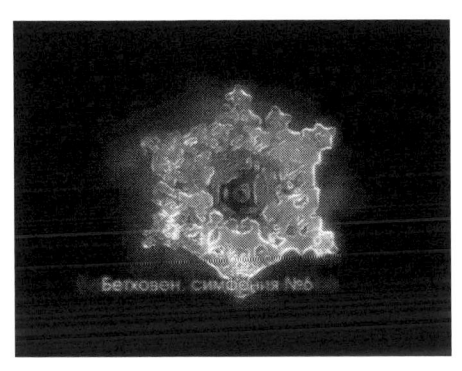

© Office Masuru Emoto, LLC
Water played the music, *Beethoven's Symphony No. 6*

© Office Masuru Emoto, LLC
Water played *Hard Rock* music

As you know, around 80 percent of our body is water. The quality of the water depends on what you may be listening to, talking about, what you are thinking, etc. The water's quality can change the quality of your health and life. Calm or aggressive behavior can be induced by the water consumed. For example, statistics show after hard rock concerts, that there is a significant increase in fights and brutality, which does not occur after a classical music concert.

Water Consumption Requirements

Our bodies need approximately two liters of water per day in the cooler months and four liters of water per day in the summer—which equates to approximately one glass of water every ninety minutes or so.

Getting Well . . . Keeping Well 155

If someone is not drinking adequate quantities of water, their body will begin to recycle fluid from the intestines to carry out these various metabolic processes. That water is essentially toxic, and individuals who don't drink adequate water can end up with symptoms, such as yellow skin tone or headaches. Thus, clean water sources and frequent clear urination, although seemingly inconvenient, is actually a sign of health as toxins are leaving the body. By the same token, frequent emptying of the intestines after eating is also a sign of excellent health. Beginning as a child, my father's sister drank a great deal of water—four to five liters daily. It seemed she had more energy and vitality than anyone. Even today at eighty-four years old, she has more energy than most people over thirty. She remains in perfect health without medical intervention.

Finding a Clean Water Source

It is challenging to find water sources that are still clean as in Switzerland or Armenia. I have never been to Switzerland; however, I am from Armenia and can attest to the condition of the water as it comes from the mountains, in the form of pure melted snow—delicious and icy cold. If you don't live in a place with pure, clean water, I recommend that you buy water from another source or learn how to purify it.

One of the simple ways to improve the condition of your water is to keep it in an uncovered glass decanter. Many undesirable elements including chlorine will evaporate, particularly if the water is in the presence of sunlight. Also, when kept in a silver decanter, water takes on antibiotic properties.

Oxygen-Supplemented Water—Added Benefits!

Locating water that has been supplemented with oxygen is particularly beneficial as it is both rejuvenating and provides a vehicle for oxygen to reach the digestive and various other organs. For example, oxygenated water reaching the stomach can heal a peptic ulcer. Oxygen reaching the interior of the body within water makes the stomach healthier, stronger, and more active, and can also travel further and increase the body's endurance in athletic competition. This belief was acknowledged in Russia, as former Soviet Union athletes were given oxygenated water in preparation for competitions. Oxygenated water can occasionally be found in health food stores or on the Internet.

Other possibilities are to get a portable oxygen tank similar to those that patients who need supplemental oxygen have delivered to their homes. Simply bubble the oxygen into a glass of water. One of my students who has followed this approach also likes to breathe the oxygen that escapes into the air around the water glass. As the oxygen is delivered from the rubber tubing attached to her tank into the bubbling water, she breathes some of the airborne concentrated oxygen going into the glass! She was amazed at how quickly the discomfort in her duodenum, which I had sensed, rapidly disappeared after a few morning glasses of oxygenated water on an empty stomach. The duodenum is the link between the stomach and the intestine where ulcers frequently form.

One other very simplistic way to get oxygen into your water is to fill a glass with water and pour the water through the air repeatedly into a second empty glass and vice versa. As the water passes through the air, it will be enriched with free oxygen. The higher one glass is from the other, the farther the water will fall before it reaches the glass and the more oxygen it will collect.

Again a Little about Water. Did You Know . . . ?

As I cited previously, water is so critical and essential to sustain life. However, we can actually live without food for a very long time. I knew of a number of people who had fasted for over forty-five days eating nothing and drinking only water. Without water one can only survive three or four days. Water is a required component of many chemical reactions that take place in the body that allow it to perform its many remarkable tasks.

One of the roles of water and oxygen is to make electro-energy for the physical movements that our brain commands. The brain uses this energy derived from water to send messages to the other organs. These electrical impulses travel across the nerves like messages on the early telegraph. These impulses are produced by a basic chemical reaction, the main component of which is water.

So the point behind this minichemistry lesson is that in order to generate electrical energy in our body, it needs a sufficient supply of water.

My Old-World Diet

There are always so many diet fads emerging. I too have tried a variety of eating styles and have settled on a few simple approaches. If your body is healthy and tolerant, you can actually digest virtually anything. When balanced and in tune, your body will use its intelligence and use everything it needs that comes in and eliminate anything unhealthy. I believe it's better to enjoy your favorite foods on occasion and make sure to detoxify periodically.

My Breakfast

Everyone is different. My day begins with honey and nuts. Both of them are ecologically very clean products. Flower nectar is considered the "food of the divine" from which honey is frequently derived. It is important to ensure one is consuming real honey as opposed to some substitutes produced from sugar processing or those that have been otherwise altered and lost their potency. To be on the safe side, I buy crystallized honey versus a liquid form. Liquidated honey may indicate that it was melted or otherwise altered as part of the bottling process. To be certain that the honey has retained its health benefits, you will notice that it will crystallize when placed in the refrigerator or when stored on the shelf during the cooler months. This is a simple test to identify a safe and desirable source of honey.

With a few teaspoons of my "heavenly honey," I mix about half of a teacup of walnuts. It's delicious and gives a great boost of energy.

I also enjoy incorporating the Armenian bread known as *lavash*. Good lavash is paper-thin as there is no true "middle," as in most soft breads. I suggest avoiding these breads as, when squeezed, they become similar to clay. Toasting and burning bread in general is not a healthy practice as you can liken it to the stomach being filled with ashes!

I also butter the lavash bread with only real butter (versus margarine) and then spread the honey and walnuts and finally roll it up into the shape of a tube. Yum!

As an alternative to honey, one can start the day with eating only fruits and vegetables.

Lunch Tips

At lunchtime, I believe in enjoying whatever foods I desire. My digestive system is hard and will be able to process it. One important guideline to know, however, is that for proper digestion, there are some nutrients that need to be thoroughly saturated in saliva as a first step before reaching the stomach. This means that one needs to eat slowly and chew very well—I chew approximately twenty to twenty-two times for each bite of food. Again, I believe in using only butter as a food-preparation oil versus vegetable oil. Vegetable oil as a living substance is in fact very health promoting when it remains raw and unheated. After cooking vegetable oils beyond 105 degrees Fahrenheit or 43 degrees Celsius, they are essentially now "dead" substances. In this form, they produce toxinlike effects in the body. So instead, I feel good about eating butter. Even the ancient yoga teachings encourage its consumption, especially in cold weather.

Dinner and Yogurt

At dinnertime, I always incorporate plain yogurt into the meal. Yogurt is a very unique product—a living food with microorganisms that are still present and delivering functional benefits in your body, including restoring the balance of your stomach and intestinal flora. These favorable bacteria begin to digest deposits on the walls and toxins that your bodily systems are challenged to remove without support. Individuals who are exposed to toxins and who supplement with yogurt, still thrive while others do not. This yogurt must be free of pasteurization as this form of heat, along with other unnatural preparation processes including a possibly irradiation, will all destroy the benefits of the yogurt. Real yogurt will go sour when placed outside of the refrigerator. Many store-bought yogurts, because of the pasteurization process, do not "turn" when left on the counter for several days. This is because their beneficial bacteria have been eliminated.

Another fascinating fact is that there is a small nation of people in the Caucasus Mountains, the Abkhazians, whose life spans are among the longest in the world, and one of the primary components of their diet is yogurt! An unusually large percentage of the inhabitants easily live from 100-120 years of age and many longer!

I consume as many fruits and vegetables as possible so that my weight and lipid levels will remain in balance. They provide the building blocks our body needs for growth and health through its life journey. However, life energy or *prana* is rejuvenated primarily through other sources outside of food intake.

Ancient teachings say that the energy that God *blew into man's body* is what changed him from the dust of the ground to a living soul.

Perhaps that may be why *breathing* is linked so strongly to our bioenergy supply.

Maintaining Prana or Life Energy, Also Referred to as Gold Energy, Cosmic Energy . . . Bioenergy

Without prana as it is called in India, or "life energy" (which we are also referring to synonymously as bioenergy and cosmic energy), it is as though there is no electricity to power a light bulb. No light equals no life. A depletion of this energy leads to a failure to thrive. With the complete absence of life energy for a moment, life ceases.

Life energy can be derived from breathing and from consuming living and very fresh fruits and vegetables. Each fruit or vegetable has a different "energetic" life span after being picked. For example, tangerines live only four hours. Beyond that, they still taste delicious; however, their life energy is depleted. Apples retain their life energy the longest—about one month. Ideally we should pick our own fruit as we don't know how long the produce has been in storage when we receive it at the market.

In the winter months when it is difficult to find any fresh fruit, the only reliable source of life energy through food is attained through seeds. Seeds can remain in a dormant condition for an indefinite period of time, and still maintain their connection with the cosmic energy field. When immersed in water, they will "awaken" and begin to take in cosmic energy.

I'd like to share an amazing story with you about seeds. Armenia is a very old country, and we celebrated our capital city Yerevan's 2,750th anniversary when I was still in Armenia. As a young student in my high school days, the Armenian government began excavations of ancient Yerevan, which at that time was known as Erebuny. They uncovered ruins of a fortress, which included structures containing multiple rooms. In one of these rooms, they found barrels of barley seeds used two to three thousand years ago to produce beer. In another room, they found a jar of honey that had become hard and transparent. Believe it or not, when they immersed these twenty-seven-century-old barley seeds in water, the seeds "woke up" and began to grow! Imagine what they can do in our bodies!

Physical Health and Yoga

The ancient Hindu discipline of yoga is primarily targeted at training one's consciousness for a state of perfect spiritual insight and tranquility. The system of specific yoga exercises and positions controls and improves both the body and the mind. For example, the yoga position referred to as *cobra* improves kidney function. The *lion* position will promote glandular health and prevents conditions, such as tonsillitis. The *Sirshasana* or *headstand* position has many benefits. It relaxes and improves the function of the heart, normalizes blood circulation, activates and calms overall organ functions and more.

There is no shortage of information about the relationship between physical exercise and its importance in the maintenance of a healthy body. Exercise is important to retain joint mobility, regulate the circulation of blood, and it ensures the even distribution of bioenergy necessary for the proper functioning of our organs, lymphatic system, and our energy centers that we spoke of known as *chakras*.

Whereas normal weightlifting exercises exclusively using artificial weights can wear out muscle and may create a heavy load on the heart, yoga exercises utilize only the weight of your body. In yoga, the positions or "asanas" are considered static or stationary physical exercises. On the other hand, the martial arts, such as kung fu and karate, incorporate movements that are more dynamic. Tai chi is thought to have more of an effect on the energetic versus the physical body. Even so, all of these modalities can contribute to achieving our goal of bringing the body and soul into harmony.

I believe yoga exercises are the most effective and easiest way to reach this state of complete balance. The specific exercises that I will recommend are easy to master at any age. *Hatha yoga* is my favorite of the yoga systems. The word *hatha*, derived from the "sun and moon," considers the opposing positive and negative energies running through our bodies and is, therefore, designed to achieve equilibrium primarily through the physical development and refined control of the body. *Hatha yoga* includes three types of exercises known as asanas, pranayamas, and mudras. The main goal of asanas is to strengthen muscles and joints, improve mobility and flexibility, resulting in an increase in blood circulation to critical parts of our body. Pranayamas are breathing exercises. The word *pranayama* comes from *prana* (breath of life) and *ayama* (broadening). Mudras are essentially asanas used specifically in meditation. The primary goal of mudras is to keep the endocrine (hormonal) glands healthy. While both adults and children can use asanas and pranayamas, mudras are not recommended until boys are at least fourteen years of age and girls have begun menstruating. Mudras should be taught by a highly experienced yoga guru, as they are not intended for physical development of the body, instead they are targeted at the endocrine system. Therefore, they are safe for those individuals who have passed the point of puberty so as not to create any interference with those naturally unfolding processes

As I mentioned, yoga combines the role of thought with physical actions to achieve a healthy body. The exercises teach you how to activate the control of the brain to achieve the desired effect on one or more designated parts of the body. The yoga literature is full of discussions about the healing powers of the various asanas and mudras. For instance, one simple position known as the *Sarvangasana* ("candle" or shoulder-stand position) is said to eradicate virtually all illnesses from indigestion, acidity, spleen and kidney problems, asthma, tuberculosis, leprosy, hemorrhoids, excessive nervousness, displacement of uterus, diabetes, and so on—a very diverse and seemingly unrelated group of ailments. The same is said of the Udiyana Bandha mudra, which when practiced is said to be capable of producing a state of ideal health and eliminating virtually any disease.

This may sound like a silver-bullet approach, and in reality, good health is achieved through a combination of many factors and contributors. Thus, a complete healing may require the practice of these positions and at the same time address some of the other modalities that I've mentioned. Illness usually develops not because of the failure of one single gland or one organ but instead it is often a group of related glands or organs that have weakened and/or contributed to an overall malfunctioning process. For this reason, we must consider the health of the entire body and use a combination of exercises and other techniques to achieve good health.

To grasp the idea of the therapeutic effect of asanas, each one consists of a unique static position for the individual to hold steady for a specific time. The period of time depends on our breathing. As we perform each asana, we will notice that our blood circulation increases at a specific bending point of the body created by that particular asana. This increase in circulation tends to "wash" or promote a healthier and more balanced state in that region of the body.

It is generally recommended that yoga be practiced twice a day—ideally sunrise and sunset. I have followed this routine and remained in good health without the necessity of any medication.

Contemplate making a commitment to take charge of your health and consider including the use of yoga pranayamas and *hatha yoga* asanas into your regimen.

The Beginning of Every Day: A Routine to Follow

Even before you perform any regular pranayama or yoga, consider starting your day with this preliminary routine:

Upon Waking

1. Open your eyes.
2. While remaining in bed, and with your hands at your side, take a deep breath, and raise your hands straight up above your head.
3. Then take a deep breath and holding it without exerting any pressure, roll to your left side and then to the right side repeatedly until you are ready to exhale. Then, exhale as you are on your right side and roll up and out of bed.
5. Repeat this exercise three to four times and, on the final time, as you turn on your right side, get up out of bed.

Upon Rising

1. Empty your bladder.
2. Wash your hands and face.
3. Clean your tongue.
   Open your mouth and push your tongue out as far as possible and beginning as far back as possible, rake your tongue of any debris. This can be accomplished in several ways: (a) using a plastic or metal tongue cleaner that can be obtained at a health food store, (b) using a teaspoon, or (c) using your three middle fingers/nails.

4. Rinse your nasal passages.
   Directions:

   a.) Prepare one glass of warm water; add a level teaspoon of salt and a touch of baking soda.
   b.) Mix it well and pour it into a deep plate or shallow bowl.
   Note: There is also a nasal rinsing pot, known as a "neti pot," that looks similar to a small porcelain teapot and is available in health food stores. It is very convenient for this purpose.
   Beginning with your left hand, hold whatever container you choose over the sink.
   c.) Using a finger from your right hand, cover the right nostril and using your left nostril, suck the water into your nasal passages until the water eventually reaches your mouth.
   d.) Spit the water out.
   e.) Reverse hand positions and do same thing, this time covering the left nostril and breathing in with the right nostril.
   f.) Repeat this process until the water in the plate or receptacle is gone.
   g.) After finishing, it will be very helpful to perform the *Sarvangasana* pose (candle position). This will further clear the nasal passages and remove any remaining water from the nostrils.

5. Drink your first glass of water.
   We must start each day by drinking water slowly (not in gulps) and the water should not be too cold—room temperature. Try to obtain the cleanest possible source of water as it will be the most useful for your body systems. For example, distilled spring water from a protected source is perfect.

6. Sun salutation.
   After drinking your first glass of water for the morning, your internal organs are still in a "sleep state" and need to be activated by doing an exercise called the sun salutation, which consists of twelve yoga positions.
   Directions:

   a.) Position yourself in an eastward direction.
   b.) Study and refer to the following diagram of poses that make up the sun salutation.

1   2   3   4

5   6

7   8

9   10   11   12

Getting Well . . . Keeping Well 165

c.) Take a deep breath and begin performing the Sun Salutations as shown in the above images. Continue to breathe fully while performing the poses.

d.) At the completion of the Sun Salutation, take a deep breath and begin exhaling while making the following four sounds. Use one-fourth of the breath you are exhaling for each. The sounds are the letters: A-O-U-M (ah oh oo um) These sounds will act to "wake up" your internal organs to start the day.

e.) The vibration of the "A" is felt in the abdominal area, the "O" in the chest, the "U" in the throat and the "M" in the head. These four sounds activate your internal organs. AOUM is also a Sanskrit prayer of the name of God. So we start our day and our yoga practice with the sound of the word *God*.

Opening and Loosening Your Joints

After the "salute to the sun," you are ready for some joint mobilization warm-up exercises. All of these exercises are done from a standing position and, unless otherwise noted, should be performed a minimum of four times on each side of the body (left and right).

Ankle Rotations

Directions: stand straight, relax, and raise your left leg until your hip and thigh are parallel to the floor. Hold it for a moment and then begin to rotate your left foot clockwise a minimum of four times. Then rotate your left foot counterclockwise a minimum of four times. Now repeat this exercise with your right leg.

Knee Rotations

Directions: stand straight, relax, raise your left leg off the floor as though you were about to start marching. While staying balanced, with your leg off the ground, rotate the part of the leg that is below the knee at least eight times clockwise then at least eight times counterclockwise. Now repeat this exercise with your right leg.

Forward Leg Swings

Directions: while steadying yourself by holding onto a fixed object, such as a wall bench, or railing, slowly swing your entire left leg forward and backward a minimum of eight times. Now repeat this exercise with your right leg.

Side Leg Lifts

Directions: using the same object to steady you, turn ninety degrees so that you are now facing and holding onto that object. Raise your left leg to the side up as close to parallel with your hip as is comfortable and then lower it again. Perform this movement a minimum of four times. Now repeat this exercise with your right leg.

Hand Stretch

Directions: stand straight and relax; with your palms facing up, and your forearms horizontal and parallel with the ground, keep a very slight bend in the elbows. Then straighten, spread, and stretch your fingers, slowly tighten your fingers into a fist and then slowly open them again. Repeat this stretching and tightening motion eight to ten times.

Wrist Rotations

Directions: while in the same position as exercise 5 (hand stretch), now make a loose fist with both hands. Slowly begin to rotate them at the wrist for a minimum of eight times clockwise, and then eight times counterclockwise.

Wrist Bend (Up and Down)

Directions: while in the same position as exercise 6, beginning with your hands facing up and forearms parallel to the floor, stretch open your fingers. Then turn your palms so that they face down, and slowly begin to bend the wrists facing upward and then downward a minimum of eight times.

Wrist Bend (Left and Right)

Directions: while in the same position as exercise 7, now bend the wrists left and right a minimum of eight times.

Forearm Rotation (Forward and Backward)

Directions: from a standing, relaxed position, hold both arms straight out to the side, at shoulder level. From the elbow, rotate the forearms in a circular motion eight times—forward and then backward.

Arm Rotation (Forward and Backward)

Directions: from the same position as exercise 9, now rotate your entire arms from the shoulders, in a circular motion—eight times in each direction—forward and then backward.

Neck Stretch (Forward—Backward)

Directions: beginning from a straight and relaxed position, slowly move your head gently forward toward your chest and back toward your spine. Repeat this for a minimum of eight times.

Neck Stretch (Left—Right)

Directions: from the same position as exercise 11, gently turn your head to the left and right a minimum of eight times.

Neck Rotation (Left—Right)

Directions: from the same position as exercise 12, slowly drop your chin toward your chest and gently rotate your head in a circular motion, first counterclockwise, toward the left and making a full circle with your head rolling gently back toward your spine until it reaches the right side. Then begin the circular rotation again starting from the right and moving clockwise toward the left. Perform this motion for a minimum of eight times in each direction.

Forward Neck Roll

Directions: from the same position as exercise 13, bend your head forward and down so you are looking toward the ground with your chin near your chest. Move your head slowly from one side to the other as if you were rolling an egg on a plate with your nose. Perform this motion for a minimum of eight times on each side.

Congratulations, your warm-up is completed! Now your body is ready for the most important step—the yoga breathing exercises known as pranayamas and the controlled health-enhancing yoga positions known as asanas.

They are very easy to master and as mentioned are intended to specifically develop the control of the brain over the health and vitality of the body. So in addition to replicating the mechanics of the exercise, we are also doing these in a relaxed meditative way so as to receive the holistic benefits.

Pranayama—Yoga Breathing Techniques

Pranayama, yoga breathing techniques, are an integral part of improving your health. In yoga, we learn that performing pranayamas can prolong life and move us closer to the attainment of physical and spiritual perfection. These exercises train us to breathe evenly and slowly and actually increase our lung volume; perhaps more importantly, they teach as to control our breathing. Yoga teachings state that God has given each of us a limited supply of air, and thus, we can live only as this supply lasts. So by controlling our breathing and learning how to breathe more slowly, we may live longer. Perhaps this may explain how some yogis have lived for four hundred years. So I recommend learning these exercises well, and practicing them frequently!

As I mentioned earlier, the word *pranayama* comes from *prana* (breath of life) and *ayama* (broadening). Pranayamas are a very important method or "art," which raises the energy level of the heart and lungs. This increase in the lung capacity allows us to maintain an adequate supply of air, and ensures sufficient oxygen to the red blood cells. Our red blood cells are responsible for collecting oxygen from the air and distributing it through the blood stream to all of our vital organs.

Our lungs are very cleverly designed organs, which like pumps continuously suck in air, distribute it to designated parts of the body, and then release the unused components back into the atmosphere. The lungs accomplish this by using three groups of muscles that both increase the volume of the lungs while inhaling, and decrease that volume while exhaling. It is important that we learn to control these three groups of muscles separately: the lower lobe, the middle lobe, and the upper lobe. Each muscle group is capable of working independently. When breathing with only the lower part of lungs, it is called "abdominal or diaphragmatic breathing." When breathing with only the middle part, it is known as "midlung breathing," and with only the upper part of lungs, it is referred to as "chest breathing." In order to master the proper yoga technique of breathing using the pranayama exercises, it is necessary to know how to control all three areas, and to be able to perform breathing that isolates the abdomen, midlung, or chest. After learning to control all three lung areas, I want to remind you that the most effective way is to breathe is using the "abdominal breathing" method.

There are several levels of pranayamas. We will talk about some of the simplest forms of pranayamas, from which anyone can benefit. Even individuals who are weak or ill will likely be able to perform these breathing techniques with relative ease and feel improvement.

Prepranayama: Learning How to Fully Breathe!

If possible, stand in front of a mirror as you learn these exercises. Prepare for each of these exercises by completely emptying your lungs.

Abdominal Breathing

In this exercise, you will try to breathe using only the lowest portion of your lungs.
Directions:

1. From a standing position, relax the abdomen and let it protrude.
2. Place one hand on the extended abdomen.
3. Exhale completely.
4. Begin inhaling, slowly feeling the air entering your lungs and filling only your abdomen. (Nothing should move except the muscles of the abdomen.)
5. When the abdomen is inflated completely, begin to exhale slowly. (It is very important to regularly use the muscles of the lower part of your lungs because this "abdominal breathing" allows the entire surface area of the lungs to receive oxygen, unlike midlung or chest breathing.)

After you feel you have mastered this type of inhalation, perform this breathing exercise eight to ten times.

Midlung Breathing

In this exercise, you will breathe using only the middle part of your lungs at the level of the solar plexus. Usually the muscles responsible for this part of the lungs are the weakest of the breathing muscles as they tend to be underutilized.

Directions:

1. Stand straight in front of a mirror and place the fingers of each hand on the left and right sides of bottom of the rib cage,—at the level where the stomach is located.
2. Exhale completely.
3. Then, without allowing the lower abdomen to inflate, inhale, filling only the middle part of lungs with air.
4. You will know that you are doing this correctly as you will feel the ribs where your hands are resting will expand outward.
5. After you feel you have mastered this type of inhalation, perform this breathing exercise eight to ten times.

Chest Breathing

Directions:

1. While remaining in front of the mirror, take a breath and allow only the chest area to inflate.
2. Place your hands on the right and left lower rib cage to monitor your focus on the chest area versus the midlung.
3. Make a conscious effort not to inflate the middle lung, which would be evident by the feeling of the ribs expanding outward.
4. After you feel you have mastered this type of inhalation, perform this breathing exercise eight to ten times.

Once you feel that you are able to control all three sections of the lungs and can breathe using only the abdomen, midlung, or chest individually, you are ready to move on to the next step. You will now combine all three sections of the lungs and breathe fully and completely. In yoga, this is called "full breathing."

Full Breathing

We can divide *full breathing* into four steps:

1. From a straight, standing position (you may wish to use a mirror), relax, exhale fully, and place your hand on your abdomen.

2. First ensuring that your diaphragm is completely deflated, begin inhaling, so that the lower part of your lungs are filling with air.
3. Once the lower part is filled, place your fingers on the end of your ribs and start to fill the middle part of your lungs. Again, your fingers will be pushed out to the side as the middle part of your lungs successfully inflates. This process should take no more than two seconds.
4. When you feel the middle part of your lungs is full of air, begin breathing through your chest to fill the upper part of your lungs. Usually the muscles of the chest are stronger, so this should be easier to do.

Note: You should already feel all three sections of your lungs filled with air. There is actually still one more section, a fourth one, which you must attempt to fill—and that's not as easy to do! This is the very, very upper end of the lungs. They are a little higher than the point of connection of the bronchi, or breathing passageways from the lungs. To fill this area with air, follow these steps:

5. With the first three sections of your lungs filled, in order to fill the fourth section, you will now actually displace some of the air by pushing it upward from the inside-lower part of the abdomen. This movement creates a pressure from the lowest part of the lung and pushes up and displaces the air that we inhaled to that upper lung area. After completing this inhalation, it is necessary to hold your breath for several seconds before beginning to exhale.
6. Exhalation begins from the lowest point in the abdomen, followed by the middle and upper parts of the lung. You also remove the air from the fourth section as well—the very upper reaches of the lungs.
7. To exhale that remaining air, it is necessary to bend your head forward, until your jaw rests on your chest. Then stick out your tongue and exhale vigorously in one short burst—as if to spit out the last bit of air.

Once you manage to put all these steps together, congratulate yourself! For the first time in your life you have fully breathed, which means your body, for the first time, got as much oxygen as it needs.

Many people today breathe from the chest almost exclusively, i.e., with the upper part of their lungs. For example, virtually 100 percent of women breathe strictly with their chests—the possible exception being opera singers, who are trained to breathe properly. Even among men, 65-70 percent breathe only with the chest. The minority of men, who do breathe more fully, have a better chance of exceeding their life expectancy.

Practice the "full breathing" for several days, until it becomes second nature and you can do it effortlessly. Then you are ready to begin learning and practicing the breathing exercises or pranayamas.

Pranayama Exercises

There are many pranayama exercises that can be found in any *hatha yoga* book. The exercises I want to offer you are among my favorites. I have practiced them for more than thirty-five years and am maintaining vibrant health. Ideally, you should perform these exercises every morning, and I recommend four iterations of each exercise.

Opening lung cells

This is the first and most important exercise and is very easy to perform. It opens all the lung cells that had been blocked as a result of air pollution, smog, bacterial infections, viruses, and other illnesses, including asthma, as an example.

This exercise must be performed as a prelude to performing any series of pranayama breathing exercises.

Directions:

1. Begin in a standing position. Relax and begin to inhale fully. As you are inhaling, tap your fingers all over your chest.
2. Then increase the force by gently pounding your chest with your fists.
3. Once your lungs are full, continue to increase the force of your fists on your chest while maintaining the air in your lungs, however, not under pressure. *This can be accomplished by keeping the throat open. (It is important to protect the heart and other organs by minimizing the pressure.)*
4. When you are ready to exhale, do so quickly and with intensity.

5. Then do cleansing breathing. (See explanation that follows on how to perform this.)
6. Repeat this exercise for a total of four times.

Cleansing breathing

Using cleansing breathing is an efficient and easy way to clean your lungs after being exposed to poor air quality whether indoors in a "sick" building or outdoors in smog or pollution. Even if you have been in an area with poor air quality for as many as thirty minutes, you can still clear your lungs with this simple exercise.

Directions:

1. Keeping your back straight in either a standing or sitting position, begin to inhale and fill your lungs.
2. After your lungs are full, hold your breath for a few seconds.
3. Purse your lips as though you are going to whistle.
4. Then exhale in pressurized spurts by pushing air through this small opening in your lips repetitively and rapidly.
5. When you have reached your last bit of air, stick your tongue out in the direction of your chest and exhale, pushing the remainder of the air out with the additional force.
6. Each time you perform a pranayama breathing exercise, follow it with a cleansing breath.

Broadening of lungs

This exercise broadens your lungs and keeps them strong and healthy.
Directions:

1. Keeping your back straight in either a standing or sitting position, exhale.
2. Then begin to inhale and fill your lungs and at the same time lift your arms, palms down, until they are parallel to the floor.
3. Hold the air in your lungs, however, not under pressure. *This can be accomplished by keeping the throat open. (It is important to protect the heart and other organs by minimizing the pressure.)*

4. Now continue to raise your arms up over your head and stretch and push them out to the side in a diagonal position and behind you for three or four times.
5. Repeat this broadening of the lungs exercise as described "of maintaining the air in your lungs and extending your arms repeatedly up and behind you in a diagonal fashion" for a total of three to four times.
6. Be sure to perform cleansing breathing between each repetition.

Squeezing of the lungs

This exercise strengthens the lung muscles.
Directions:

1. From a standing position, place the left hand under the left armpit on the side of the rib cage and the right hand under the right armpit under the rib cage. The thumbs should be pointing back and the four fingers to the front.
2. With your arms relaxed, begin to inhale while resting your hands gently on the sides of your rib cage.
3. When your lungs are full, hold your breath without pressure (throat open) for two seconds.
4. Then begin to exhale and, at the same time, press on the sides of the ribs to help push the air out completely. This sound will be like a tire deflating.
5. Then take another deep breath and perform cleansing breathing.

Nerve calming exercise

This exercise will calm your nervous system.
Directions:

1. Ideally, from a standing position, with a straight back, begin to inhale and at the same time raise your arms, with your palms up until they are parallel to the floor.
2. While you are holding your breath without pressure, bend your arms at your elbows and bring your palms forward toward your face, and then squeeze your fists.

3. Then allow your arms to return to the parallel position and release your fists.
4. Repeat this motion four times while maintaining the air in your lungs.
5. Perform this exercise for four repetitions.

Isolated abdominal breathing

This is a very simple and easy to do exercise that is considered so powerful according to the teachings of yoga that it is said to cure one hundred incurable illnesses. It strengthens the abdominal muscles and increases oxygenation. It is one of the quickest ways to eliminate abdominal fat too!

Directions:

1. From either a standing or seated position, begin by exhaling deeply.
2. Then inhale and exhale rapidly from the abdominal area, yet at a comfortable level, approximately two times per second.
3. Continue this until you wish to stop.
4. Then perform a cleansing breath.
5. Perform this exercise for four repetitions.

Breathing in the "lion position"

For this pranayama, the ideal position is the lotus, which is a more advanced pose. For the beginner, any seated position is fine. This position is helpful to the thyroid gland.

Directions:

1. Begin in a seated or lotus position.
2. Inhale deeply.
3. Lower your head and place your chin on your chest.
4. Allow your tongue to fall from your mouth and in the direction of your chest.
5. Begin to exhale and create a tone that sounds like a lion—thus the lion position!
6. Perform a cleansing breath.
7. Perform this exercise for four repetitions.

"Moon and sun" breathing

This breathing technique isolates the flow of air through the left and right nostrils.

Breathing through the left nostril is known as *moon* or negative polarity breathing. Breathing through the right nostril is known as *sun* or positive polarity breathing.

This exercise will teach you how to sense and activate your sympathetic nervous system (controls fight or flight responses) and parasympathetic nervous system (controls resting activities and digestion. You may wish to refer to an anatomical diagram that will show you the location of these two systems. Nerves from each of these systems traverse either side of the spine. They are the conduit through which we receive cosmic energy (bioenergy). The strength, activity and energetic cleanliness of these autonomic nervous systems will impact how much cosmic energy (bioenergy) reaches your nerves, concentrates in the chakras (energy centers) and then travels to the related organ systems. This exercise will help to maximize this efficiency. (This is discussed in more detail under the exercises for working with the energetic centers—chakras.)

Directions:

1. From a seated position, place your middle finger of your right hand gently over the nose so that the fingertip rests between the eyes on the bridge of your nose.
2. Place your right thumb over the right nostril so that it impedes the flow of air.
3. Rest the right little finger above the left nostril gently so air can still pass.
4. Begin to inhale slowly and deeply so that the airflow is isolated to the left nostril. Imagine it coming through the nostril and around the right side of the back of the head and down the right side of the spine. This is the side of the spine that controls the nervous system.
5. Once your lungs are full of air, hold your breath without pressure for several seconds.
6. Then cover the left nostril with the little (pinky) finger and release the thumb from the right nostril.
7. Imagine the energy now coming back up the left side of the spine, around the back of the head to the right and then exhale fully out of the right nostril.

8. Pause for a few seconds before inhaling again, this time through only the right nostril, by keeping the left nostril covered. Imagine the energy coming through the right nostril around to the back of the head and down the left side of the spine, which controls the nervous system.
9. Once your lungs are full of air, hold your breath without pressure for several seconds.
10. Then cover the right nostril with the thumb and release the little (pinky) finger from the left nostril.
11. Imagine the energy now returning back up the right side of the spine, around the back of the head to the left and then exhale fully out of the left nostril.
12. Perform a cleansing breath.
13. Perform this exercise for four repetitions

Note: It is important that you use your right hand for this exercise. The thumb represents positive polarity and, therefore, must cover the right nostril. The little (pinky) finger represents negative polarity and, therefore, must cover the left nostril in order to maintain that polarity. If you were to use your left hand, the thumb (or positive polarity) would naturally fall on the left nostril, which would neutralize the negative polarity of that nostril with the positive polarity of a thumb. Conversely the little pinky finger of the left hand which carries a negative polarity would rest on the right nostril and would, therefore, neutralize the positive polarity of that nostril. Why is this important? Because it would neutralize the bioenergy flow, which is what we are trying to revitalize through this exercise.

Asanas

If you are familiar with hatha yoga, you may know there are hundreds and hundreds of asanas (or poses). It would be very challenging to learn all of them. I have reviewed a number of books on the subject of *hatha yoga*, and my favorite was written by Swami Vivekananda entitled *The Secret Side of Yoga* and translated into Russian. The book I had was published in St. Petersburg in 1914. I have been unable to find an English translation, so if you do, let me know. It is like "gold!" In the meantime, I will share some of his information with you. I have been practicing these asanas for over thirty-five years. I have maintained excellent, drug-free health. Living in Los Angeles, I have visited many highly recommended yoga schools. One area in which the schools can improve is to demonstrate that, in fact, yoga is more than just an aerobic or other exercise program. It is actually a mental and meditative practice with emphasis on proper control of prana or cosmic energy. This is an important aspect that should be taught to and by our yoga instructors.

To cite an example of how yoga asanas need to be more fully integrated as part of a larger spiritual practice, we can consider the common asana known as the *corpse pose*. This pose is performed by lying on your back and completely relaxing. This is often the extent to which this pose is explained and taught in most yoga schools. However, this is really just the beginning of the intention and "power" of this pose. It is one of the most important poses as it also allows for the integration of all of the poses that preceded it so that the student will grow in their practice and realize the physical and spiritual benefits. Furthermore, once in this relaxed and supine position, you can work very effectively with bringing life energy into your energetic system.

Yoga is in fact a very powerful and brilliant science about the human body, teaching us the secrets to maintaining strength and long-term health. There are many books written on this subject as well as resources you can find on the Internet and elsewhere that will provide examples of the asanas. Your yoga practice should begin and conclude with the asana known as the Shavasana or *corpse pose*. Beyond the physical aspects of assuming these poses, I would like to explain the energetic aspects of a yoga practice.

Shavasana-Asana or Corpse Pose

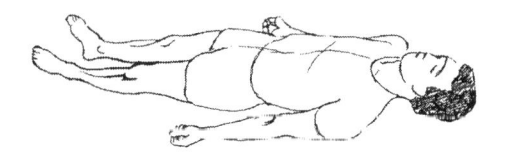

Directions:

1. Lie down on your back with your legs slightly spread at approximately 35 to 40 degrees and with your arms straight at your side and your palms facing upward. You have the option of keeping your eyes open or closed.
2. Relax and begin to "breathe" into your body parts one at a time. I Begin with the left foot and continue up the leg. Imagine your left

foot is "breathing in" gold cosmic energy that has the brilliance of a halo. Imagine that if there is any "dirt" or negative energy inside your foot that this gold energy will be brought in to replace it.
3. This gold cosmic energy "cleans and shines" the foot, and with each exhale, deposits any unwanted "debris" or negative energy outside of the body. Continue with this visualization from the foot, up each part of the leg. As you master this visualization exercise, you will actually begin to feel the sensation like "wind" as the gold cosmic energy passes through your leg.

Some individuals can feel the presence of the gold energy immediately. Others may detect it within as few as three days of practice or sometimes over the course of possibly several months. Be patient, as you too will be able to feel the movement of cosmic energy through your body temple.

4. Perform this cleansing visualization exercise three to four times and then go to the right foot and proceed up the right leg.
5. After cleansing the right leg with cosmic energy, you will move upward to the reproductive organs.
6. This time the gold energy will travel in between your legs and into the reproductive organs, i.e., the uterus and ovaries for women, and the penis and prostate gland for men. This gold cosmic energy will rejuvenate male sexual abilities and stamina to a more youthful state.
7. Perform this action three or four times and then continue up to the abdomen.
8. Imagine the gold energy passing through an opening that reaches and "cleans" the organs, including the stomach, pancreas, intestines, spleen, liver, and gall bladder.
9. After performing this "cleansing" three or four times, begin again, this time entering from your back. Feel the gold energy passing through your back and reaching the same organs now from this new angle. This time you are also reaching and cleansing the kidneys.
10. Perform this activity three or four times. Then continue upward to the chest.
11. Imagine this bright gold cosmic life energy traveling up your chest and nourishing the heart and lungs from both the front and back sides of the chest.

12. Then move your visualization to the left and right arms, bringing the gold energy in three to four times on each side.
13. Now continue to the thyroid. This is a very important gland as it regulates metabolism.
14. Next go to your head and imagine it as a large round "sponge." See the life force gold energy entering through the pores in the "sponge." Imagine your brain is cleansed with this gold energy. Now you have cleaned and cleared each body part individually.
15. Finally bring everything together and imagine the body as a whole as a large "sponge," breathing in the gold life force energy across the entire body. Imagine it entering through all of the pores in the sponge, and each time you perform the exercise, imagine that there are even more entry points appearing for gold energy to penetrate this "sponge."

This entire meditative exercise should take approximately ten minutes.

Once you have completed this meditative visualization, you are ready to begin mastering your yoga asanas. The asanas as mentioned have both a physical benefit of squeezing the muscles, enhancing the speed and efficiency of blood flow to the organ systems, as well energetically cleansing the body as described. When you are performing the asanas, also meditate on the fact that the poses simultaneously propel life energy to their focus areas. These areas and the surrounding organ systems are cleansed and restored with this shining "gold energy."

Below are a series for you to experience. Begin each day with them, and you will be amazed at how your health, flexibility, quality of life, and longevity will be positively impacted.

Morning Asanas

For each asana listed below (A-I), I recommend that you perform them for a minimum of three repetitions and in the provided sequence. For each one, follow the four-step procedure outlined that will integrate the energetic cleansing aspect. You will want to have your diagram of the chakra system handy (Refer to Figure 1 on page 142) as each pose will be related to a particular chakra. You will also be guided to visualize an opening and closing lotus flower, see Figure 2 that follows:

Figure 2 Closed and Open Lotus Flowers

1. Imagine each chakra as a lotus with its petals closed and oriented perpendicularly to your spinal column.
2. On the first repetition of the asana, use your prana (cosmic energy / bioenergy) to fill the chakra then simply rest and feel a sense of calm.
3. On the second repetition, imagine sending the prana there and also cleansing the lotus and its petals.
4. Finally, on the third repetition, send the prana to the chakra and imagine the lotus radiating the gold cosmic energy on its own.

A. *Utthan Padasana*—The Leg Lifting Pose

While performing this asana and the four step process three times as described above, send the energy to the first chakra known as the *root* chakra.

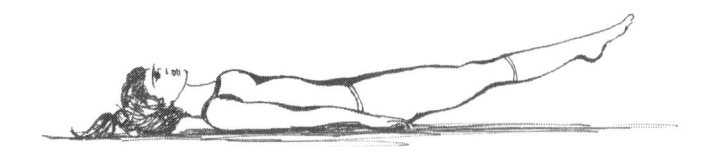

B. *Pashimotanasana*—The Noble or Powerful Pose

In this asana, follow the same process and send energy to the next chakra known as the *sacral* chakra for each of the three repetitions.

C. *Bhujangasana*—The Cobra Pose

Send energy to the *will* chakra using the same process

D. *Salabhasana*—The Locust Pose

Send energy to the *solar plexus* chakra using the same process.

E. *Sarvangasana*—The Shoulder Stand Pose

Send energy to the *heart* chakra using the same process. Remember, this is the best exercise for the heart.

F. *Matsyasana*—The Fish Pose

Send energy to the *throat* chakra using the same process.

G. *Trikonasana*—The Triangle Pose

Send energy to the *listening* chakra using the same process.

H. *Dhanurasana*—The Bow Pose

Send energy to the *third eye* chakra using the same process.

I. *Halasana*—The Plow Pose

Send energy to the *crown* chakra using the same process.

1. After completing the above nine poses, perform the Shavasana (or corpse pose) again.

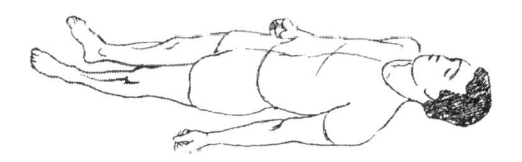

This time, you can lie down on a floor, relax and start to breathe in gold energy using your entire body. Do this three to four times.

Evening Asa

At evening time, I recommend the following asanas in this sequence:

1. *Shavasana*—The Corpse Pose

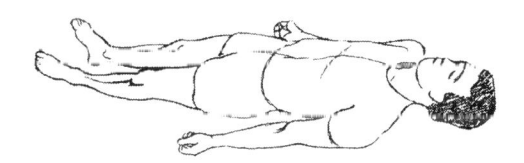

2. *Sirshasana*—The Headstand Pose

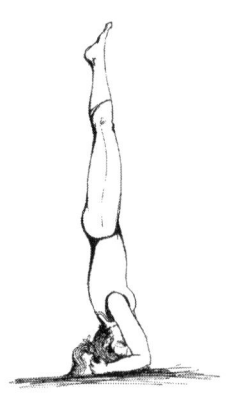

3. *Sarvangasana*—The Shoulder Stand Pose

4. *Matsyasana*—The Fish Pose

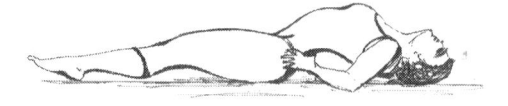

5. *Pashimotanasana*—The Noble or Powerful Pose

6. Bhujangasana—The Cobra Pose

7. *Salabhasana*—The Locust Pose

8. *Dhanurasana*—The Bow Pose

9. *Halasana*—The Plow Pose

After completing these evening asanas, we will end with *Shavasana*.

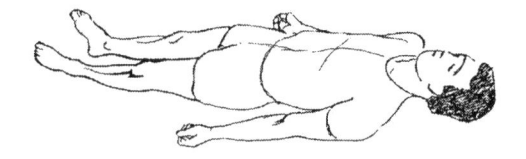

Spiritual Cleansing of the Chakras

Over time, once you have mastered the asanas and have realized the physical benefits of this practice, you may wish to learn how to clean the chakras *spiritually* using the following guidelines.

1. Begin by requesting the presence of a spiritual figure with whom you find a connection. Then ask for their help in opening and cleansing your chakras.
2. Lie down on your back, relax, and take several deep breaths. After you are fully relaxed, you will cleanse each of your nine chakras one at a time, beginning by breathing into each chakra. You can reference the figure on page 142 for their locations.
3. For each chakra, you will follow a similar pattern. You will begin by inhaling and putting your attention on the location of that chakra. Imagine when you inhale that the chakra is being filled with bright gold life energy or prana. Imagine that the chakra looks like a closed lotus. Relax that chakra.
4. Then with the help of your spiritual master, pray for assistance in cleansing and opening your chakras. As you inhale, send prana to each chakra and imagine that energy cleanses the gold lotus and opens all the petals.
5. Next, for each chakra, inhale and imagine that the chakra represented in your mind's eye as a lotus flower, will begin to radiate like a miniature sun and then rotate 90 degrees. Finally, ask your holy spiritual master for the following spiritual gifts associated with each chakra as follows:

    1. Ask to receive *divine wisdom*, which is the spiritual side of the *first* or *crown chakra which physically relates to consciousness*.

2. Ask to receive *divine contemplation*, which is the spiritual side of the *second* or *sight chakra*.
3. Ask to receive *divine understanding*, which is the spiritual side of the *third* or *listening chakra*.
4. Ask to receive *divine glorification*, which is the spiritual side of the *fourth* or *throat chakra, which physically relates to the power of words*.
5. Ask to receive *divine love*, which is the spiritual side of the *fifth* or *heart chakra*.
6. Ask to receive *divine harmony and beauty*, which is the spiritual side of the *sixth* or *solar plexus chakra*, also known as the *chakra of sensing*.
7. Ask to receive *divine will*, which is the spiritual side with the *seventh* or *will chakra*.
8. Ask to be aligned with *divinity or godliness*, which is the spiritual side of the *eighth* or *sacral (passion) chakra*.
9. Ask for the *Holy Spirit* to inspire you, which is the spiritual side with the *ninth* or *root chakra*.

Upon completing the spiritual cleansing of the chakras, they become very active and fully energized.

10. Therefore, you can now relax, and as we learned previously, fill your overall body as if it were a sponge, with this gold energy using this "body breathing" technique.
11. Then envision all nine chakras in the open lotus position and pointing upward.

Congratulations!

We've covered a lot of ground in these two appendices in terms of breathing, water, diet, yoga, controlling bioenergy, and more. Now you have the beginnings for some new ideas and practices to incorporate into your daily routine. I suggest spending at least 30 to 40 minutes in the mornings and evenings using the various physical activities and lifestyle recommendations that I have explained.

Please feel free to contact me about how these lifestyle practices and the use of, bioenergy have made a difference in your life and the lives around you or if I can answer any questions and support your healing journey info@HealngWithBioenergy.com

Appendix C

More of the Science That Supports the Concept of Bioenergy

To further your interest, I wanted to touch on some of the other science that has evolved and is being researched that can help to explain how this entire phenomenon of bioenergy.

The Evolution of the Fundamentals of Physics Provides Evidence of Bioenergy

If we look at how the laws of physics have evolved and become better understood over several hundred years, the evidence of bioenergy is clear.

After discovering the basic laws of mechanical physics, Sir Isaac Newton experienced some difficulty with certain aspects of optics. His first paper on the subject, *The Light and a New Theory about Colors,* was introduced to the Royal Academy in London in 1672. He talked about the idea that light has weight. He raised the fact that light reveals itself in some instances as a stream of elementary particles and in other situations presents itself with wavelike characteristics. This dynamic nature of light was later validated by further experimentation. However, even so, the laws of Newtonian physics did not have the ability to explain the phenomena of electromagnetism and electromagnetic fields, and it forced science to look elsewhere for an explanation of these concepts.

As a result, Maxwell's four differential equations emerged. These four equations could be considered the foundation of *new physics*. Without them, much of today's scientific progress would have been impossible.

These equations consider the fact that light has a constant speed in the environment, which is demonstrated in the following equation:

$$C = 7.5 \times 10 \text{ in /sec}$$

This became the basis of Einstein's Special Theory of Relativity, which was published in 1905. According to this theory, there is nothing in nature that has a speed higher than the speed of light. It resulted in the idea that such constants as weight and time became dependent on speed. Time stopped being independent and having only one direction. Einstein's theory gave us the entrance into the worlds of atomic and laser energy. If not for this progress from Newtonian physics, the biggest achievement of humanity would be the train!

Later, physics took another new step forward. In 1964, the famous physicist G. C. Bell published his theory about the interrelationship of interatomic elementary particles. His theory proved mathematically that these interactions in fact do not depend on time and space. The smallest changes in the particle immediately influence other particles without any time passing. This theory was also experimentally proven. Contrary to Einstein's theory, it demonstrated that speed has no limits and has an infinite potential.

This explains why a patient could be on the other side of the galaxy and still be affected by my bioenergetic treatment!

Unlike physics, modern medicine has not yet found its "Maxwell equations" and still remains in the Newtonian world. Today, although it readily uses many of the world's great scientific discoveries to create new tools for medical diagnosis and treatment, these methods and resultant limited treatment success are not questioned or challenged to a degree that seems to warrant further development and perfecting.

A human being is still not looked upon as the complex creature that it is—consisting of more than just its physical body—made up of an astral, an ethereal, and a mental body surrounded by a biomagnetic field. However, modern medicine has finally acknowledged the fact that the human body is controlled by electrical impulses sent from the brain. In a popular book, *The Human Body*, John O. E. Clark writes,

> *Our nervous system is functioning with the help of weak electrical signals, or impulses passing through them. With great speed they transfer the information from sensitive nervous endings*

to the brain and send back an order to respond in a fraction of a second.

Physics confirms it. Electrical impulses control and operate all the organs in the body. A very complicated chemical reaction provides the energy necessary.

The main source of energy for us is a substance known as adenosine triphosphate (ATP). When it combines with water, it divides in two subcomponents, adenosine diphosphate (ADP) and adenosine monophosphate (AMP). As a result, an electrical current is produced. The chemical formula looks like this:

$$ATP + H2O \longrightarrow ADP + AMP$$

In spite of these facts, modern medicine refuses to widely recognize the presence of a bioelectromagnetic field around the human body.

A well-known law of physics states that *if an electrical charge or current runs through a wire (conductor), an electromagnetic field is formed around the wire.* Every second, there are numerous processes taking place in the human body. Our brain is constantly transmitting electrical impulses to various destinations in our body using a large quantity of electromagnetic fields. The full complement of these individual microfields creates an electromagnetic field around our body. So it stands to reason that the effective functioning of this "bioenergy" field and its component "wiring," or meridians, would have an effect on the vitality of an individual.

Aura and Energy Field Hypotheses

Today in Russia, there are two schools of thought surrounding the concept of aura studies. One school believes that we are dealing with a combination of two well-known phenomena: electrical and magnetic fields, which, when combined, alter the resultant energy quality and appearance. The second believes that an aura is a completely separate phenomenon connected with another still-unidentified physical field. Scientists believe that every form of energy, including the aura field, carries within it a variety of information vectors, most of which we are currently incapable of deciphering.

Our individual biofield or aura contains the complete information about us. The ability exists within human beings to see the aura field in its full-color range as a clairvoyant can demonstrate. However, most of us have not been able to tap into those abilities. I personally know of only two people who can see the full-spectrum aura.

Scientists hypothesize that there are, at a minimum, five levels of energy in nature and perhaps as many as seven to nine. The first level is the kinetic energy movement, which has been recognized since the 18th century. The second is the electromagnetic field, which was discovered in the 19th century. Now we are on the verge of the third level, which is a biological energy field. Some scientists believe that an informational field will emerge as the fourth level. It is too early to speculate about the concept of the higher levels that are connected with the forces controlling the universe.

INDEX

A

abdominal pain, 74. *See also* Mihran (abdominal pain patient)
acupuncturist, 90
Adan, 72
ADP (adenosine diphosphate), 193
adrenal tumor, 62–63. *See also* Tomas (adrenal tumor patient)
Agababov, Sergay, 65
Albert (kidney pain patient), 47–48
Alpik (lymphatic cancer patient), 44, 46
Amatuny, 70–74
America, 58, 79, 81, 109, 113, 115
AMP (adenosine monophosphate), 193
Anahid (friend with breast tumor), 52–53, 57, 106–7
Anahit, 128–29
Anna (child with neurological disorder), 49–52
Anush. *See* Manukyan, Anushavan
Arbat Street, 81–82
Arevik, 125
Arkadi, 80–81
Arkady, 35, 35–36
Armen (basketball player), 32–33
Armen (child with asthma), 49
Armen (esophageal tumor patient), 89–90
Armenia, 33, 38–39, 62–63, 66, 69, 128, 156
Artsrun, 50–52
Arvanov, Artick, 62–64
asanas, 114, 161–64, 166, 168, 178–79, 181–82, 185, 188
Asmik (child with spleen disorder), 80–81
asthma, 49. *See also* Armen (child with asthma)
ATP (adenosine triphosphate), 193
auras
  biofield, 21–22, 26–29, 116–17, 137–44
  energy fields, 13, 28, 138
  hypotheses, 193
Azura, 98–99

B

Barstow Hospital, 126
basketball player. *See* Armen (basketball player)
bioenergy

body's capacity for, 122
evidence of, 26–27, 29–30
healing through, 109
immune system effects of, 90
physics proof of, 191–93
psychic surgery techniques of, 71–73
Soviet interest in, 67–69, 96
bioenergy healing
for low blood platelet levels, 109–10
mentor in the study of, 38–39
protecting self during, 43–46
testing anxiety through, 100
use of, 110–12
bioenergy therapy, 58, 119–20
with hypnosis, 54–56, 99–100
on liver disorders, 87–88, 103–5
bioenergy treatments, 29, 65, 117
benefiting from someone else's, 86–87
effectiveness of, 110–11
processes in, 42–43, 65
biofield, 21–22, 24, 26–27, 64, 116–17, 133, 147, 194
manipulation of, 28
meditative access to, 28–29
programming the, 140–41, 143–44
sensing objects in, 138–40
sensing of, 137–38
blood platelet levels, 109–10. *See also* Shogik (patient with low blood platelet levels)
boyfriend of patient with tumor on uterus. *See* Vladimir
brain tumors, caused by cell phones, 107
breast cancer, 6, 90, 111, 125, 203
breast tumor, 52–53, 57, 90–91, 106–7. *See also* Anahid (friend with breast tumor); Robin (breast tumor patient)

breathing
abdominal, 169–70
chest, 169, 171
cleansing, 174–75
full, 150–51, 171, 173
"moon and sun," 177–78
stomach, 170
Burbank Hospital, 96

C

cancer
chance of recovery from, 58–59
healing exceptions of, 41
need for new approach to, 113–15
psychological impacts of, 113, 115
surgery for, 105–9
treatments for, 107–9, 115
Cedars-Sinai Hospital, 103, 106
chakras, 38, 121, 145–48, 188–89
healthy, 145
nine-chakra system, 146
chemotherapy, 13, 56–60, 91, 106–10, 115–16, 119
Chumak, Allan, 110
Clark, John O. E.
Human Body, The, 192
Cliburn, Van, 32–33
colon cancer, 101–2. *See also* Margaret (colon cancer patient)
comatose, 99. *See also* Cyrus (comatose child); Jennifer (comatose patient)
Conversations (Karabekov), 61
CT scan, 30
Cyrus (comatose child), 99

D

daughters. *See* Arevik; Sona
Davidashvily, Juna, 68
diet, 110, 117, 150, 158–59
digestion, 159, 177
Dilijan, 59–60
diseases, 26, 85–86, 115–16, 120, 122
distance healing, 37, 39, 69, 93, 110–11, 133. *See also* Garnik (distance-healing patient)
duodenum, 157

E

Emoto, Masuru, 152–55
endocrine system, 25, 162
energy
  effects of obstruction of, 116–17
  levels of, 194
  power of residual, 30
energy channels, 46, 55, 105, 116–19
  causes of obstruction in, 116–17
  obstructed, 26, 41, 117
energy field
  experiment, 28
  hypotheses, 193
energy scan, 64, 69, 74, 97
esophageal tumor, 89–90. *See also* Armen (esophageal tumor patient)
ethereal body, 100, 114, 120–22, 192
exercise
  for calming the nerve, 177
  for the heart, 183

F

family friend. *See* Anahit
father of child with neurological disorder. *See* Artsrun

forest, 45
fruits, 110, 158, 160

G

Garnik (distance-healing patient), 110–11
Gevork, 87–88
Gilayvan, Yuri, 125–26
Good Morning America, 24
Gretchen (patient with testing anxiety), 100–101
Guillen, Michael, 24
Gurdjiev, George Ivanovich, 39
guru. *See* Arkady

H

headaches, 82–83, 85, 103, 107–8
head tumors, 101–2. *See also* Robert (head tumors patient)
healer. *See* Adan
healing, 29, 70–71, 120
  remote, 37, 91–111
heart, 61, 96–97, 161, 169
herbs, 90
honey, 158, 161
House of Scientists, 44
human body, 121–22
Human Body, The (Clark), 192
Human Energy Systems Lab, 24
husband of lymphatic cancer patient. *See* Vardan
hypnosis, 32–35
  with pranayama, 35

I

immune system, 90–91, 114–16, 118–20, 122–23
Inessa, 50–52
Ivazovsky, Ivan
  Ninth Tidal Wave, The, 76

J

Jennifer (comatose patient), 96–100, 126
Julia (neck tumor patient), 106

K

Kane, Elizabeth, 93
Karabekov, Rudik, 65
  Conversations, 61
Keep Your Breasts (Moss), 111
KGB, 68, 80
KGB representative. *See* Arkadi
kidney pain, 47–48. *See also* Albert (kidney pain patient)
kidney problem, 76–77. *See also* Tital (kidney problem patient)
kidneys, 62, 76–77, 109, 180
Kiev, 54, 62–63
Kirlian Effect, 22–23
  experiments, 22–23
Kremlin Hospital, 44, 59, 62–63, 71, 82–83

L

Laert (boy with neck tumor), 56–58, 106–9
Las Vegas, 125
lavash, 158

Lena (liver cancer patient), 103–6
liver cancer, 87, 103–6. *See also* Lena (liver cancer patient)
liver disorders, 87–88
Lord's Prayer, The: An Esoteric Study (Steiner), 122
Los Angeles, 46, 85–87, 93, 96, 103, 129, 178
lungs, 37, 61, 126, 151, 169
lupus, 61, 64–66, 110. *See also* Zara (lupus patient)
lymphatic cancer, 44, 46. *See also* Alpik (lymphatic cancer patient); Seda (lymphatic cancer patient)
lymphatic system, 44, 102, 114

M

Manukyan, Anushavan, 54, 62–64, 110, 129
  car accident of, 125–27
  collaboration with Dr. Amatuny of, 73–75
  exposure to hypnosis of, 33–35
  as a healer, 48, 52–54, 56–58, 81, 83, 87–88, 90
  healing practice in the United States of, 85
  introduction to yoga of, 35–36
  learning to heal of, 36–37
  moving to America of, 79–83
  in a newspaper article on bioenergy healing, 96
  postcoma of, 127–30
Manukyan, Elada, 19, 33, 50, 52, 60, 76, 95, 99, 118
  breast cancer of, 41–43, 46
  car accident, 125
  healing her husband, 126–27

Margaret (colon cancer patient), 101–2
meditation, 28
Mepisashvily, Gia, 38–39, 43, 67–69
Mesmer, Franz Anton, 26
metastasis, 118
Mihran (abdominal pain patient), 74
mobility exercises
   ankle rotations, 166
   arm rotation, 168
   forearm rotation, 167–68
   forward leg swings, 166
   forward neck roll, 168
   hand stretch, 167
   knee rotation, 166
   neck rotation, 168
   neck stretch, 168
   side leg lifts, 167
   wrist bend, 167
   wrist rotation, 167
morphine, 107–9
Moscow, 35, 44, 61–63, 67–68, 71–73
Moss, Susan
   Keep Your Breasts, 111–12
mother of child with neurological disorder. *See* Inessa
Motor Vehicles Department, 76–77
mudras, 161–62
multiple sclerosis, 93–94. *See also* Randy (multiple sclerosis patient)
muscles
   elimination of pain in the, 133
   weakest of the, 170

N

Natasha (uterine tumor patient), 86
National Institutes of Health, 24–25
neck tumor, 106. *See also* Julia (neck tumor patient); Laert (boy with neck tumor)
negative energy
   accumulation of, 27–28, 116–17
   being responsive to bioenergy, 28–30
   elimination of, 36, 42
nerve calming exercises, 175
neurological disorder, 49–52. *See also* Anna (child with neurological disorder)
Newton, Isaac, 191
Ninth Tidal Wave, The (Ivazovsky), 76

O

osteoporosis, 61, 64–65
OVIR (emigration service), 75, 79–80

P

pain relief, 47, 76, 133
Persian patient. *See* Azura
"phantom leaf" experiment, 23
Philippines, 71, 73
plasmabiological body, 23
poses. *See* asanas
prana, 131, 160, 162, 169, 182, 188
pranayamas, 35–36, 150, 161–63, 168 70, 173, 176
prednisone, 109–10
professor. *See* Amatuny
prostate cancer, 91–93. *See also* Vahktang (prostate cancer patient)
psoriasis, 86–88

R

radiotherapy, 58, 110, 115–16, 119
Randy (multiple sclerosis patient), 93–94
relativity, theory of, 192
Robert (head tumors patient), 101–2
Robin (breast tumor patient), 90–91
Rubik, Beverly, 24

S

San Bernardino, 129
San Diego, 98–99
Sarnoff, Galya, 96–97
Seda (lymphatic cancer patient), 58–61
self-healing, 109, 120
Shahinyan, Edward, 75–76
Shogik (patient with low blood platelet levels), 109–10
skepticism, 82, 85
Sona, 33, 41
son of woman with liver cancer. *See* Gevork
Soviet Union, 23, 28, 65, 156. *See also* USSR
spine, 50, 55, 64, 121, 145, 168, 177–78
spleen, 80–81, 109, 162
spleen disorder, 80–81. *See also* Asmik (child with spleen disorder)
stabbing, 36–37. *See also* Tigran (stabbing patient)
Steiner, Rudolf
  The Lord's Prayer: An Esoteric Study, 122
sun salutation, 164, 166
surgeon. *See* Amatuny
surgery
  counterproductiveness of, 117, 120
  psychic, 71–73

T

Tbilisi, 38, 67–68
testing anxiety, 100–101. *See also* Gretchen (patient with testing anxiety)
Tigran (stabbing patient), 36–37
Tital (kidney problem patient), 76–77
Tomas (adrenal tumor patient), 62–63
toxins, 156, 159
transcendentalism, 38
transcendental medicine, 38
tumors, 43, 69–70
  as accumulated negative energy, 27–28
  adrenal, 62
  benign, 118
  brain, 107
  cancerous, 56, 116
  development of, 116–17
  esophageal, 89
  kidney, 62–63
  malignant, 56–58
  neck, 56–58, 106
  softening of, 27, 43, 87, 106, 119

U

ulcer, 156–57
University of Southern California (USC), 114
USSR, 32, 68, 74, 80–81
uterine tumor, 86. *See also* Natasha (uterine tumor patient)

V

Vahktang (prostate cancer patient), 91–93
Vardan, 60
vegetables, 110, 158, 160
Vladimir, 86–87

W

walnuts, 63, 158
water
  consumption requirements of, 155
  generating electrical energy through, 157, 193
  molecular structure of, 151
  oxygenated, 156
  sources of clean, 156, 164
willpower, 133–35

X

x-rays, 24, 30, 65, 87, 103, 107

Y

Yerevan, 33–34, 49–50, 61, 66, 69, 160
Yerevan Drama Theater, 34, 36, 128
Yerevan Medical Institute, 70

yoga
  asanas in, 114, 178, 181
  exercises of, 36, 161–62
  hatha yoga, 161, 178
  physical health benefits of, 36, 132, 162–63
  pranayamas, 36, 169
yoga poses
  corpse, 179, 185
  lion, 161, 176
  lotus, 176
  pharaoh, 137–38, 141–43, 148
yogis, 44–45, 150–51, 169
yogurt, 159

Z

Zara (lupus patient), 61, 64–66, 110
Zero Institutes, 68–69

Biographical Information and Basis for Book

Anushavan Manukyan was born in 1948 in Yerevan, the capital of Armenia. In 1972 he completed his university studies as a physicist specializing in quantum optics. He began working at the Armenian Institute of Physics, while at the same time, teaching high school physics. In 1974 he became the first Yoga teacher in Armenia and opened a school, and in 1981, after mastering karate, he became one of the first karate teachers in Yerevan and taught in several schools around the area. In 1980 he met with Gia Mepisashvily, who introduced him to bioenergy therapy techniques as a healing modality. While teaching how to cure many illnesses using this technique Gia believed that the treatment of cancer was not possible using this protocol. In fact, it was considered dangerous to the practitioner, who could absorb the associated negative energy and manifest that very illness.

Some time later, Anushavan's wife received a diagnosis of breast cancer, so Anushavan was compelled to use these techniques on her, regardless of the warnings from his teacher. As a result, he broke through the perceived barrier and discovered that bioenergy therapy had the potential to cure anything, including cancer. In particular, he discovered that if the patient *had not* undergone surgery, radiation therapy or chemotherapy it was his experience that she had more than a 97 percent chance to respond. In fact, the side effects of surgery and radiation affected the body's energetic healing system so dramatically that it made it very challenging, and sometimes impossible, to reverse the effects.

In 1988 Anushavan left Armenia with his wife Elada and two daughters, Sona and Arevik, and moved to Los Angeles. He became the first person in the United States to actually receive a license as a bioenergy therapist—and it was because of his request and powerful demonstration of his skills that such a license came into being. He brought a new healing modality from his home in Armenia to many people in the U.S., reversing a vast array of illnesses, in particular, cancer. Through his continued practice he discovered that the bioenergy therapy worked even remotely and even at a very long distance. So now proximity was no longer a requirement. He "saw patients" from England, France, Russia—all over!

Later, he himself was injured in a serious car accident and his wife used the very same bioenergy techniques she had witnessed to help him survive and recover from a coma when doctors provided no hope of normal life.

This universal energy field is available to all of us. The author's dream is for all people to know that they have healing abilities that go beyond what medical institutions can offer.

Anusshavan's book encourages readers to find their own healing paths and master their innate self-healing abilities as he did. We can all learn basic foundational approaches to a healthy life style and how to use the power of the bioenergy field to support our minds, bodies, and souls and our journeys through life.

Anushavan Manukyan practices and teaches bioenergy healing in Los Angeles.